THE BASES OF
MODERN SCIENCE

By

J. W. N. SULLIVAN

First published in 1928

J. W. N. SULLIVAN

John William Navin Sullivan was a popular writer on science and literature, author of a celebrated study on Beethoven. He wrote one of the earliest accounts of Einstein's *General Theory of Relativity* and circulated amongst the literary elite of London, including Aldous Huxley, Wyndham Lewis and T.S. Eliot. Sullivan's early days are somewhat murky however. He was known to frequently lie about his upbringing, telling Aldous Huxley that he was born in Ireland and attended *Maynooth* with James Joyce. In fact, he was born on 22 January 1886 in the East End of London, England, and worked at a Telegraph company from 1900 onwards. The directors of the firm recognised his outstanding mathematical capabilities, and financed Sullivan's study at the *Northern Polytechnic Institute*. From 1908 to 1910, Sullivan studied and researched at *University College London*, but left without a degree and moved to America, where he became a journalist. Sullivan's peripatetic life continued, and in 1913 he returned to Britain, still employed as a journalist, before working for the ambulance service in Serbia during the First World War. After the war, Sullivan wrote for *The New Witness* and *The Athenaeum*. He married his first wife, Sylvia Mannooch in 1917 and they had one daughter, Navina, born in November 1921. It was at *The Athenaeum,* one of the best known and important literary reviews of the 1920s that Sullivan was introduced to the London literary world. He continued contributing literary and scientific articles to *The Athenaeum* well into the 1920s, but also wrote for the *Times Literary Supplement, The Adelphi* and *John O'London's Weekly*. During this time, Sullivan wrote his explanation of Einstein's theory of relativity, as well as numerous articles noting the new spirit of creativity in the sciences and the possibility of their reconciliation with the arts. These articles were collated in *Aspects of Science* in 1923. Sullivan went on to pen the well-received study, *Beethoven: His Spiritual Development* (1927) and contributed to *An Outline of Modern Knowledge* (1931). Sullivan separated from his first wife in 1921, and married Vere Bartrick Baker in 1928, with whom he had a son, Navin. In the early 1930s, Sullivan was increasingly troubled by poor health however, and was diagnosed in 1934 as suffering from *disseminated sclerosis,* a form of creeping paralysis. He died on 11 August, 1937 in Chobham, Surrey.

THE BASES OF MODERN SCIENCE

BY

J. W. N. SULLIVAN

CONTENTS

CHAPTER   PAGE

      PREFACE 7

I. THE FIRST SKETCH 9

II. THE NEWTONIAN CONCEPTIONS: SPACE, TIME, MASS 26

III. THE ETHER THEORY 42

IV. HEAT AND ENERGY 63

V. MOLECULES AND ATOMS 80

VI. ELECTROMAGNETISM 88

VII. THE ATOM OF ELECTRICITY 110

VIII. THE ELECTRIC THEORY OF MATTER . . . 129

IX. RELATIVITY 153

X. GEOMETRY AND PHYSICS: THE FINITE UNIVERSE 170

XI. NEW PROBLEMS 193

XII. GENERAL CONCLUSIONS 203

      APPENDIX 212

5

PREFACE

This book is an attempt to expound the main ideas of physical science in non-technical language. The scheme of the book may well seem ambitious, since I begin with the Copernican revolution and end with the new theories of the atom. But the scientific reader would speedily discover that I have ignored hosts of details and avoided many difficult doctrines. These abstentions I have practised deliberately, for I believe that I should otherwise harm my main purpose, which is to communicate, in a readily understandable way, the most essential and important scientific concepts to that large class of intelligent readers who have had no scientific training. To this end I have not once introduced a mathematical expression, however simple. I have found by experience that men of but literary education, highly intelligent, logical, imaginative, have all their mental powers instantly paralysed by the sight of a mathematical symbol. This book, which traces in outline the great scientific enterprise of describing Nature mathematically, is therefore written without mathematical formulæ. Theories, or aspects of theories, which would not lend themselves to such emasculated treatment I have therefore ignored. Nevertheless, I have not consciously omitted anything that seemed to me essential to the understanding of the scientific outlook, so far as that outlook can be understood without technicalities. And I believe that, for the sort of understanding of science that intelligent but non-scientific readers are after, technicalities are not necessary.

J. W. N. Sullivan.

Chobham, 1928.

CHAPTER I

THE FIRST SKETCH

THE scientific picture of the universe is the result of a careful and but lately developed process of selection applied to the elements of man's total experience. From our total reactions to Nature science selects a small part only as being relevant to its purpose, and the criteria of selection it adopts were not formulated till within the last few centuries. Familiar as we are with the scientific way of approach, and natural as it now seems to us, there is, in truth, nothing obvious about it. Faced with the amazing number and variety of impressions we call Nature it is only by great effort that man has learned to discriminate those that may be successfully used for constructing science. Science is the result of an attempt to bestow a certain kind of order and a certain kind of coherence upon experience. Or we may say that it is an attempt to discover that kind of order and coherence. Science is the result of a mental adventure, and there was no *a priori* reason to suppose that the adventure would be successful. Indeed, the very notion of undertaking the adventure did not occur until late in man's history. Quite other kinds of order and coherence than those now regarded as typically scientific were first sought for. Some of these are not inconsistent with the scientific ordering of experience although they do not form part of it. For instance, the notion of some mediaeval writers that animals have a symbolic significance, that the wolf is a symbol of covetousness, and so on, does not conflict with biology, although biologists do not employ it. Similarly, the notion that God created the stars to be pleasing adornments of the earth is not logically incompatible with modern astronomical knowledge, although the scale on which we now know the adornment to be carried out suggests a singular lack of artistic restraint.

What we call science is not man's first attempt to introduce order into Nature; it is merely his first attempt to introduce a particular kind of order. The mediaevalist lived in an orderly universe, a universe in some ways more orderly than our own. But the order he discovered or imported was obtained by employing quite different criteria. His chief criterion was teleological. Phenomena were ordered in accordance with their bearing on human purposes. Man was at the centre of a dependent universe. The sun existed to give him warmth and light, and rain fell to nourish his crops. This was the guiding principle in terms of which Nature was explained. Everything that existed and everything that happened was useful to man, even if only for the purpose of teaching him salutary moral lessons. Thus the fact that Nature does not appear wholly beneficent to man was not an insuperable difficulty for this method of thought.

Nature, as conceived in this way, was part of a larger scheme. The central and dominating fact was man and his immortal destiny. The "material universe" was merely a setting within which a moment of this destiny was being worked out. The ultimate reason for Nature's existence was to be found in its bearing on that destiny, for its contribution towards the final end of man, which was union with God. As a natural consequence of this point of view phenomena were explained in terms of their supposed purposes. The "why" of phenomena, not the "how" of phenomena, was the question that interested the mediaeval mind. The mediaeval universe was informed through and through with *purpose*. Men did not interpret the temporal passage of Nature as a bare succession of events, but as the passage from potentiality to actuality. All things conspired together towards some divine end. The merely spatial and temporal connections of phenomena were not considered to be of importance compared with their logical connections. Phenomena were regarded as exemplifying certain general logical principles and as serving a universal purpose. The general mediaeval outlook made the assumption that Nature was rational a reasonable one. Since both Nature and man had the same author, and Nature was designed to forward man's destiny, it was not unreasonable to suppose that the workings of Nature should proceed in a manner intelligible to the human mind. Later, when science gave up this basis for the rationality of Nature, there was nothing to replace it but a pure act of faith. Science replaced the mediaeval scheme by a different one, but the new scheme did not contain within itself any grounds for supposing that it must be successful. Science itself provides no ground, beyond the pragmatic one of success, for supposing that Nature forms an orderly and coherent whole. Science, therefore, rests not upon a rational basis, but upon an act of faith. The scientific belief in the rationality of Nature is seen to be, historically, an inheritance from a system of thought of which the other terms have been discarded.

Nevertheless, the rational character of Nature as assumed in science is quite different from the character assumed in mediaeval thought. According to the mediaevalist nature is teleological; according to science it is mathematical. The two statements are not incompatible with one another. They do not conflict, but they testify to entirely different attitudes towards phenomena. The mediaeval assumption fitted in with a completely general metaphysical and religious outlook. The scientific assumption seems to have

11

been, in part, the expression of personal predilection and, in part, of neo-Platonic philosophy. Kepler was convinced that Nature was necessarily mathematical. Newton regarded the attempt to describe Nature mathematically as an adventure that might or might not be successful.

The Copernican theory, which may be taken as the first great departure in the application of the mathematical criterion, had nothing to recommend it but its mathematical simplicity. It appeared contrary to the immediate evidence of the senses, and those who insisted that knowledge must be empirical were bound to reject it. Also, it left unanswered various grave objections, such as that objects on the surface of the earth would be left behind —that is, would leave the earth like projectiles—if the earth were really in rapid motion. Nevertheless, the greater mathematical harmony introduced by the theory into phenomena was sufficient to ensure its triumph. In saying this, however, we must remember that the Copernican scheme was perhaps put forward merely as a mathematical device. Copernicus's scheme, like that of Ptolemy that it replaced, was put forward as the simplest geometrical representation of celestial phenomena. It was *convenient*; the question of whether it was true or untrue was not explicitly discussed. Almost everybody agreed that the astronomical realm was fundamentally geometrical; the question was, What is the simplest system of geometry that will "save the phenomena"? Copernicus argued that if, as a mathematical device, the sun rather than the earth is made the centre of reference, the geometry is simpler. The cumbersome epicyclic system of his day offended the mathematician in Copernicus and led him to speculate whether some other scheme was possible. He hunted up the opinions of other philosophers, and found that some of the ancients had held the idea that the earth moved. In his own words:

" When from this, therefore, I had conceived its possibility, I myself also began to meditate upon the mobility of the earth. And although the opinion seemed absurd, yet because I knew the liberty had been accorded to others before me of imagining whatsoever circles they pleased to explain the phenomena of the stars, I thought I also might readily be allowed to experiment whether, by supposing the earth to have some motion, stronger demonstrations than those of the others could be found as to the revolution of the celestial sphere.

" Thus, supposing these motions which I attribute to the earth later on in this book, I found at length by much and long observation, that if the motions of the other planets were added to the rotation of the earth and calculated as for the revolution of that planet, not only the phenomena of the others followed from this, but also it so bound together both the order and magnitude of all the planets and the spheres and the heaven itself, that in no single part could one thing be altered without confusion among the other parts and in all the universe. Hence for this reason in the course of this work I have followed this system."

Whatever degree of reality Copernicus himself may have attached to his system, he put it forward as a mathematical device. As he says, " Mathematics is written for mathematicians, to whom these my labours, if I am not mistaken, will appear to contribute something."

Although the idea that Nature is susceptible of mathematical description was not prominent in the orthodox Aristotelianism of the Middle Ages, it had been enthusiastically believed by prominent thinkers before Copernicus, notably Roger Bacon and Leonardo da Vinci. And in the time of Copernicus it was one of the tenets of a vigorous neo-Platonic school in Italy, a school that directly influenced Copernicus. This conviction was founded upon certain Pythagorean teachings that we would now dismiss as mystical.

The belief that the universe lent itself to mathematical description became a much more radical conception with Kepler. To Kepler the universe was essentially mathematical. Indeed, the *cause* of phenomena being as they were was that they fulfilled certain mathematical relations. The Copernican hypothesis was not regarded by Kepler as a mathematical device. Its mathematical simplicity and aesthetic charm was the proof that it was true; it was, moreover, the *cause* that it was true. Thus when Kepler found that the " five regular solids " could be inserted between the spheres of the six planets known to him, he considered he had found the cause of the planets being six in number. This relation which, even if it really existed, we should now regard as puerile, was quite as important to Kepler as his three laws of planetary motion. As he says: " The intense pleasure I have received from this discovery can never be told in words. I regretted no more the time wasted; I tired of no labour; I shunned no toil of reckoning, days and nights spent in calculation, until I could see whether my hypothesis would agree with the orbits of Copernicus or whether my joy was to vanish into air."

The procedure on this occasion was typical of Kepler. He was convinced, on *a priori* grounds, that a sufficiently profound study of the astronomical system would reveal the mathematical harmonies on which it was based. It so happened that, amongst the large number of mathematical relations he found, he found some that were fruitful. But Kepler himself seems to have had no criterion of fruitfulness. In this his attitude was not typical of the modern scientific mind. But in another respect he was entirely modern, in his insistence on empirical confirmation. Fantastic as were some of Kepler's published " discoveries," they were merely the few survivors of a host of fantasies that had been killed by the facts. Kepler, one of the most luxuriant and wildly theoretical of *a priorists,* nevertheless constantly appealed to the facts of observation and unhesitatingly abode by their verdict. That the cause of phenomena was their embodiment of mathematical harmonies he had no doubt, but only observation could decide which, of all possible mathematical harmonies, were those actually embodied in Nature. Kepler thus escaped the chief danger to which his type is subject, the danger of underestimating the need for empirical knowledge. The Greeks had not escaped this danger, and Kepler was certainly as extreme a believer in the mystical significance of mathematics as was any Pythagorean. He not only believed that Nature was essentially mathematical, but he was also convinced that the only certain knowledge we could have of the world was knowledge of its quantitative features. And this was true owing to the very structure of the human mind, which was such that it could know nothing completely except quantities.

Kepler is sometimes referred to as being the first man to exhibit the modern fully scientific mind. This is not quite true. Kepler believed more than is necessary. It is not a part of the scientific outlook to believe that Nature is necessarily mathematical, although physical science, as pursued hitherto, is very largely the result of a successful attempt to describe phenomena mathematically. But this attempt is undertaken as an adventure. It is not supposed that there is any *a priori* reason why the attempt should be successful. In his recognition of the need for observation, however, Kepler is certainly modern.

In estimating the importance of Kepler's work for the formation of the scientific outlook we must distinguish between mathematics and aesthetics. When Kepler replaced circles by ellipses as the orbits of the planets, he substituted mathematics for aesthetics. Both the circle and the ellipse are, of course, mathematical figures, but the circle was chosen as the path appropriate to the heavenly bodies, not for the mathematical properties it enjoys, but because of its " perfection "—a purely aesthetic criterion. Kepler was himself extremely sensitive to such considerations. He did, indeed, make eighteen successive attempts to fit circles to the planetary orbits before he definitely adopted ellipses. His faithfulness to observation overcame his aesthetic preferences. It is possible, also, that, being a real mathematician, he found the aesthetic properties of ellipses sufficiently charming. In saying that the planetary orbits were elliptical Kepler was, incidentally, opposing the same general attitude that was encountered by Galileo when he stated that there were mountains on the moon. It was objected to Galileo's teaching that, the moon being a heavenly body and therefore a perfect sphere, it could not be disfigured by mountains. The aesthetic criterion was, therefore, one of those that had to be abandoned in order that the scientific outlook could be formulated. As a matter of fact, the aesthetic criterion still plays a part in science, but it is now called " simplicity."

Galileo approached the modern scientific outlook more nearly than did Kepler. Like Kepler, he believed that Nature lent itself to mathematical description, but he does not seem to have regarded the fulfilment of mathematical relations as the *cause* of phenomena. He regarded mathematics rather as the one and only true key that would introduce order and coherence into our sense impressions. For ages men had tried in vain to make an orderly universe out of the universe of sense, and their efforts were vain because they had never learned the true alphabet of that universe. The true alphabet is the principles of mathematics. To introduce order into the universe of sense impressions, man must pay attention to the quantitative aspects of those impressions and discover the mathematical relations that exist between them. From these relations, once discovered, certain consequences can be mathematically deduced, and these consequences can then be verified by experiment. Galileo was so confident that natural phenomena proceeded according to mathematical principles that he did not himself, he tells us, always feel the need for experimental verification. Fortunately, however, the task of convincing his opponents made him undertake such verifications. But he realized quite clearly that the mathematical relations sought for must be exact. It is not sufficient, for instance, to know that bodies fall with an accelerated motion. In his own words:

" Neither doth this suffice, but it is requisite to know according to what proportion such an acceleration is made; a problem that I believe was never hitherto understood by any philosopher or mathematician, although philosophers, and particularly the peripatetics, have writ great and entire volumes touching motion."

But in this task of introducing order into our sense impressions by the mathematical key it sometimes happens that we reach results that seem incompatible with the direct testimony of our senses. An instance is provided by the Copernican system. Faced with this problem, Galileo plumps for reason. He says:

" I cannot sufficiently admire the eminence of those men's wits, that have received and held it to be true, and with the sprightliness of their judgments offered such violence to their own senses, as that they have been able to prefer that which their reason dictated to them, to that which sensible experiments represented most manifestly to the contrary."

It is interesting to compare this passage with one in a letter to Kepler:

"Oh, my dear Kepler, how I wish that we could have one hearty laugh together! Here at Padua is the principal professor of philosophy, whom I have repeatedly and urgently requested to look at the moon and planets through my glass, which he pertinaciously refuses to do. Why are you not here? What shouts of laughter we should have at this glorious folly! And to hear the professor of philosophy at Pisa labouring before the Grand Duke with logical arguments, as if with magical incantations, to charm the new planets out of the sky."

It would seem that the professor of philosophy at Pisa was doing the same thing Galileo so admired Copernicus for doing, namely, preferring that which his reason dictated to him to that which sensible experiments represented most manifestly to the contrary. It is obvious that the logical issue involved cannot be left here, and that the transition from the world of sense to the " real " world of mathematical order has not been satisfactorily achieved by Galileo. Galileo has not, in fact, satisfactorily and explicitly told us what are the characteristics of his ": real " world. We gain our knowledge of it through our senses, and yet it may contradict the testimony of our senses. What he assumes may perhaps be expressed by saying that, of all our perceptions, there are some only that are perceptions of real existents in the real world. These are the perceptions between which mathematical relations may be found to exist. Other perceptions are not perceptions of actual existents. It need not be denied that they are correlated with actual existents but, if so, the actual existents are of a different nature from the perceptions. Thus, Galileo supposed that our perceptions of extension, motion, mass are perceptions of real existents which are precisely extension, motion, mass. But our perception of heat, although not illusory, is not a perception of heat as a real existent. It is a perception of the real existent "motion." Other perceptions, as of colours, sounds, odours, are of this curious class. As this doctrine played a very great part in the formulation of the scientific outlook we will give Galileo's own account of it:

"But first I want to propose some examination of that which we call heat, whose generally accepted notion comes very far from the truth if my serious doubts be correct, inasmuch as it is supposed to be a true accident, affection, and quality really residing in the thing which we perceive to be heated. Nevertheless, I say, that indeed I feel myself impelled by the necessity, as soon as I conceive a piece of matter or corporeal substance, of conceiving that in its own nature it is bounded and figured in such and such a figure, that in relation to others it is large or small, that it is in this or that place, in this or that time, that it is in motion or remains at rest, that it touches or does not touch another body, that it is single, few, or many; in short by no imagination can a body be separated from such conditions: but that it must be white or red, bitter or sweet, sounding or mute, of a pleasant or unpleasant odour, I do not perceive my mind forced to acknowledge it necessarily accompanied by such conditions; so if the senses were not the escorts, perhaps the reason or the imagination by itself would never have arrived at them. Hence I think that these tastes, odours, colours, etc., on the side of the object in which they seem to exist, are nothing else than mere names, but hold their residence solely in the sensitive body; so that if the animal were removed, every such quality would be abolished and annihilated. Nevertheless, as soon as we have imposed names on them, particular and different from those of the other primary and real accidents, we induce ourselves to believe that they also exist just as truly and really as the latter. I think that by an illustration I can explain my meaning more clearly. I pass a hand, first over a marble statue, then over a living man. Concerning all the effects which come from the hand, as regards the hand itself, they are the same whether on the one or on the other object—that is, these primary accidents, namely motion and touch (for we call them by no other names)—but the animate body which suffers that operation feels various affections according to the different parts touched, and if the sole of the foot, the kneecap, or the armpit be touched, it perceives, besides the common sense of touch, another affection, to which we have given a particular name, calling it tickling. Now this affection is all ours, and does not belong to the hand at all. And it seems to me that they would greatly err who should say that the hand, besides motion and touch, possessed in itself another faculty different from those, namely the tickling faculty; so that tickling would be an accident that exists in it. A piece of paper, or a feather lightly rubbed on whatever part of our body. you wish, performs, as regards itself, everywhere the same operation, that is, movement and touch; but in us, if touched between the eyes, on the nose, and under the nostrils, it excites an almost intolerable tickling, though elsewhere it can hardly be felt at all. Now this tickling is all in us, and not in the feather, and if the animate and sensitive body be removed, it is nothing more than a mere name. Of precisely a similar and not greater existence do I believe these various qualities to be possessed, which are attributed to natural bodies, such as tastes, odours, colours, and others."

The sole characteristics of the real world, according to Galileo, are quantitative. Interaction between our bodies and the rest of the real world somehow produces in our minds those qualitative differences we naïvely attribute to the real world. He says:

20

" But that external bodies, to excite in us these tastes, these odours, and these sounds, demand other than size, figure, number, and slow or rapid motion, I do not believe; and I judge that, if the ears, the tongue, and the nostrils were taken away, the figure, the numbers, and the motions would indeed remain, but not the odours, nor the tastes nor the sounds, which, without the living animal, I do not believe are anything else than names, just as tickling is precisely nothing but a name if the armpit and the nasal membrane be removed." Galileo assumes, then, that amongst the characteristics we attribute to the real world are some that would remain " without the living animal " and some that would not. The first class of characteristics, size, motion, etc., are quantitative, and this seems to be one of the reasons why they are preferred. Galileo's assumption that they exist " without the living animal " is not justified by his arguments and, indeed, rests on grounds which were only later made explicit and which have never been made logically irrefragable.

It will be noted that, in this argument, Galileo does not mention mass amongst the characteristics of his real world. But he was quite aware of the fact that the characteristics of size, shape and motion were not sufficient to enable phenomena to be mathematically described. He was aware of the importance of *weight*. That bodies of the same size, shape, and motion could have different weights was a fact about them that one had to take into consideration in attempting a mathematical description of their behaviour. The characteristics of the real world were size, shape, weight and motion. This real world was, Galileo believed, atomic in structure. Therefore, in Galileo's view, the real world consisted of atoms possessing size, shape and 'weight, and these atoms were in motion. From their combinations the material universe arose. The atomic motions were produced by *force*, but on the ultimate nature of force Galileo, with a reserve most extraordinary in his time, refused to speculate.

In this view of the world which, in its main lines, has been the scientific view for centuries, the notion of cause has undergone a profound transformation, and the notions of space and time have acquired an importance they never before possessed. On the pre-scientific view the cause of a phenomenon was found by asking *why* it occurred. Thus some bodies fell downwards and some others, as flames, soared upwards, *because* each body tended towards its natural place. For Galileo the cause of a phenomenon was found by asking *how* it occurred. The list of preceding conditions which were always followed by the phenomenon in question was the cause of the phenomenon. The cause was transferred, as it were, from the end to the beginning of the process. Motion, which, on this view, was an important and fundamental characteristic of the real world, involves the notions of time and space.

Time and space, therefore, became fundamental in the new outlook. Special relations between bodies were supposed to conform to the principles of geometry—Euclidean geometry. Time, also, was regarded as something that could be exactly formulated mathematically. It was his grasp of the mathematical nature of time that led Galileo to devise more exact clocks. We have, then, the real world conceived as consisting of bodies located in space and time and possessing no characteristics but those that can be given mathematical formulation. The scientific task was thus clearly defined; it was to account for all the variety of phenomena in terms of these fundamental concepts.

The Galilean analysis of the real world has, as we have said, dominated science until quite recent times, chiefly through the influence of Isaac Newton. Newton adopted, extended, and made more explicit the fundamental Galilean concepts. But that the Galilean analysis was not inevitable is shown by the work of Descartes. Descartes was, if possible, even more convinced than Galileo that the key to the operations of Nature was to be found in mathematics. But he elaborated a theory in terms of a somewhat different set of fundamental entities. He tried to account for phenomena in terms of the geometrical property of extension and the property of motion, without importing the notion of weight or of mass. Galileo had found that the notions of extension and motion were not sufficient, by themselves, to describe the observed behaviour of bodies. He therefore imported other fundamental notions, such as weight and momentum. Obviously, therefore, the Cartesian analysis was bound to fail. Descartes overcame this difficulty by inventing an " Æther," a universal medium whose vortical motions explained those features of phenomena which could not be derived from bare extension and motion. The properties of this æther were left vague. It amounted to no more than an evasion of the problem of determining the necessary and sufficient characteristics of Nature that should allow of the complete mathematical description of the observed behaviour of bodies. Newton, with his truer instinct, recognized the insufficiency of the Cartesian conceptions, and adopted those of Galileo.

The scientific outlook was very largely due, as we have seen, to the predilections of the mathematician. The metaphysic assumed by these men—that Nature must be essentially mathematical in character—was nothing but an expression of their predilection. The harmful effects of this assumption are most apparent in the case of Descartes, who deduced erroneous laws of Nature from quite gratuitous *a priori* assumptions. Even Galileo, as we have said, was inclined to think that mathematical deductions did not require experimental confirmation. None of these men, great as they were, manifested the scientific mind in complete purity. But, together with these great mathematical legislators, there existed men of a very different type, who made their indispensable contribution to the formation of the modern scientific outlook. These were the empiricists, such men as Gilbert, Harvey, Boyle. Their characteristic note was their insistence upon the experimental investigation of particular cases. Even when they agreed with the mathematicians that phenomena could be ultimately reduced to mathematical relations between atoms in space and time, they also pointed out that a considerable control of such phenomena could be gained without making this reduction—by stopping at directly perceived qualities, several links short, as it were, of the atomic end of the chain of causation. As against the extreme mathematician's tendency to deduce all phenomena by reasoning from a few general principles, they insisted upon the importance of experience. Boyle, in particular, found that the *a priori* reasoners sometimes arrived at conclusions which were contradicted by actual experimental evidence. Nevertheless, Boyle did not doubt the general world-view of the mathematicians. For Boyle, as much as for Galileo and Descartes, the world was a mathematical machine. The chief importance of these men for modern science was as protagonists of the empirical element in science. Science, to begin with, had a tendency to be rather too mathematical. Although their assumptions and technique were different, the early mathematicians were not wholly free from the failings of the scholastic philosophers they replaced.

We may, therefore, summarize as follows the chief ingredients of the scientific outlook as it existed at the time Newton appeared. It was assumed that some of our perceptions, as of extensions, motion, weight, were perceptions of objectively existing characteristics of the real world. It was further assumed that others of our perceptions, as of colours, sounds, scents, were not perceptions of objectively existing characteristics of the real world, although they were correlated with such characteristics. This assumption was usual, but not universal. Boyle, for instance, did not hold it. The characteristics of the real world were completely definable mathematically. A complete mathematical description of the real world was the aim of science. Together with this was a strong insistence on the necessity of empirical observation. To assume certain principles and deduce their mathematical consequences was not sufficient. The results so obtained must be verified experimentally. And it was not quite clear that the characteristics so far discovered, as extension, etc., were sufficient to account for everything. Such phenomena as magnetism led to the invention of " Æthers " and " influences " of considerable obscurity. The notion of the æther was also required to support the dogma that one piece of matter could not act on another piece of matter " at a distance." Motion, particularly on the Cartesian philosophy, was regarded as communicated wholly by contact. The æther, which filled all space not occupied by matter, was the medium by which the necessary shocks were transmitted.

To make clear and definite the concepts of the mathematicians, to combine this achievement with complete adherence to the empirical method, and to vindicate the whole outlook by applying it with complete success to the description of phenomena, was the achievement of Isaac Newton.

CHAPTER II

THE NEWTONIAN CONCEPTIONS: SPACE, TIME, MASS

THE ideal aim of science is to give a complete mathematical description of phenomena in terms of the fewest principles and entities. We have to find, under the baffling variety of Nature, the essential and irreducible elements whose combinations produce that variety. The attempt can claim to be successful, so far, only for those sciences that ignore the chemical, living and mental properties of bodies. The primary concepts in terms of which the science of physics is constructed, such concepts as force, energy, mass, velocity, acceleration, momentum, etc., have to be supplemented by others in the science of chemistry and, for the sciences of life and mind, are so far from being sufficient, that they have hardly yet been found to be even relevant. The mathematical description of Nature, therefore, is possible so far only for a limited region of phenomena. Whether such a description is even theoretically possible for the whole of Nature is not a question that can be decided on this evidence. The doctrine of materialism is a doctrine that such a description is theoretically possible and, further, that the description would proceed in terms of the concepts mass, energy, acceleration, etc., that have been used in physics until recent times. This doctrine has no longer any plausibility, since the concepts in question have been found inadequate even in physics itself. But the inadequacy of the old concepts does not, of course, affect the theoretical possibility that a mathematical description of the whole of Nature may be given in terms of the new concepts or, if not in those, in terms of concepts not yet invented. The question is of no importance except that it enables us to realize the nature of our general assumption. Either we suppose that phenomena possess aspects that cannot be described mathematically, and that therefore science is a partial description of Nature, since it deals only with those aspects that are mathematical, or we suppose that the present partial description given by science is due merely to the fact that its present technique enables it to deal only with the simplest cases. That is, we may suppose that those phenomena which do lend themselves to scientific description are described exhaustively. When, for instance, Newton says:

"It seems probable to me that God in the beginning formed matter in solid, massy, hard, impenetrable, movable particles, of such sizes and figures, and with such other properties, in such proportion to space, as most conduced to the end for which he formed them; and that these primitive particles, being solids, are incomparably harder than any porous bodies compounded of them; even so very hard, as never to wear or break in pieces: no ordinary power being able to divide what God himself made one in the first creation," are we to suppose this list of properties of the ultimate particles to be exhaustive? That Newton himself thought so is very doubtful; most of his followers, however, did think so. It was this belief that nothing existed but the primitive particles, and that these particles were exhaustively described by their mathematically definable properties, that enabled Laplace to say that a sufficiently great mathematical intelligence, given the distribution of the ultimate particles in the primitive nebula, could deduce the whole future history of the world. This extraordinary act of faith was prompted by the fact that the Newtonian set of abstrac-

27

tions, space, time, mass, enabled a complete description of the phenomena of motion to be given. For a long time these three concepts, as Newton formulated them, were the cardinal concepts of the science of physics. Other concepts used in physics were derivative from them. Thus a velocity is the rate of increase of distance with time. Acceleration is the rate of increase of velocity with time. Momentum is the product of mass and velocity. And so on. In an account of modern scientific ideas, therefore, we must begin with these three cardinal conceptions.

The notion of mass is necessary, as we have said, to make possible the mathematical description of the motions of material bodies. Two bodies may be of the same geometrical size and shape and yet, under the influence of a given impact, acquire different motions. This is, of course, one of the commonest facts of experience. Golfers do not play golf with little cannon balls. But to disentangle the particular property of matter that is responsible for this difference was not easy. The property is constant for the same piece of matter; a given impact will always produce the same amount of motion in the same body. The property in question is not the same thing as the weight of the body. A body at the equator weighs less than it does at the north pole and, on a journey to the moon, it would reach a point where it weighed nothing. Nevertheless, it would always acquire the same amount of motion through the same impact. The weight of a body depends on its position relative to the centre of the earth. The property we are trying to disentangle is independent of the body's position relative to surrounding bodies.

This property Newton called *mass*. In defining mass he assumed the notion of force, and this notion has since become highly sophisticated or even dissolved away altogether. But the Newtonian notion of force was based, psychologically, on the common human experience of muscular effort. The ordinary human experience of pushing and pulling was the basis of the scientific conception of force. Bodies which " attracted " one another were vaguely supposed to be pulling at one another, and this pull was usually supposed to be transmitted through some intervening medium—an " Æther."

Now Newton stated that the effect produced by a given force, whether that force was a push or a pull, depended not only on the force but on the mass of the body on which it acted. And he gave a precise rule by which the force could be measured, whether the body was at rest or in motion. For the action of force is measured by the *change of motion* of the body. If the body is originally at rest this change of motion is, of course, the whole motion. If the body is already in motion the action of the force is to change the motion. But the change of motion, for a given force, depends on the mass of the body. The more massive the body the smaller the change in its motion for a given force. Newton stated that the force was precisely measured by taking the product of the mass and the change of motion. In speaking of change of motion we mean the change that occurs in a definite standard interval of time—say one second. Change of motion, so defined, is called acceleration. Newton states, therefore, that the force acting on a body is equal to the body's mass multiplied by its acceleration.

This celebrated law has an aspect that deserves attention. It defines mass in terms of force and force in terms of mass. The one directly observable quantity introduced is acceleration. If we start by knowing " force " we can obtain " mass " in this way. If we start by knowing mass we can obtain force. But the definition merely defines them in terms of one another. It would seem that either mass or force is a superfluous concept. We can, however, obtain the mass of a body experimentally without introducing the notion of force at all. If we allow different bodies to collide their velocities after collision are usually different from their velocities before collision. But there is a certain simple function of the velocities before collision which has the same value as that same function of the velocities after collision. This function involves " co-efficients," one for each body. These co-efficients are the masses of the bodies concerned. The layman may feel that this procedure is merely a device and not a discovery. The difference, in science, is not clear. But the merit of this device, if it can be so called, lies in its usefulness. For when, by our collision experiment, we have determined the mass of a body, we find that this mass is quite independent of that particular experiment. We get the same value for the mass of that body whatever other bodies we make it collide with. Hence we have discovered a characteristic of this body which, within the conditions of these experiments, remains invariant. Further, this characteristic remains invariant in all conditions of temperature and is quite unaffected by the presence of surrounding bodies. It has thus a high degree of permanence and is therefore important. We suppose, also, that it persists unaltered through all chemical changes. The mass we obtain from the collision experiments is called the *inertial mass* of the body. It is the mass, we may say, that is concerned in impacts between bodies.

But a body also possesses another characteristic called its *gravitational mass*. All bodies, as we know from the most famous of all scientific laws, Newton's law of gravitation, attract one another. The degree of this attraction depends upon their distances and their masses, we are told. But the "mass" mentioned here is not obviously the mass involved in the collision experiments. The mass involved in those experiments owes nothing, so far as we can see, to that mysterious power of attraction that all bodies exhibit. It would exist, we may suppose, even if bodies did not attract one another at all. The "mass" involved in gravitation appears to be, like the "charge" on electrified bodies, an entirely separate thing. We measure its amount by making "attraction" experiments, as we before made "collision" experiments. We find that it also remains invariant. And we make the astonishing discovery that the relation between a body's inertial and gravitational mass is exactly the same as for any other body. This relation is entirely independent of the physical conditions or chemical constitution of the bodies concerned. No other two properties of a body are linked together in this indissoluble way. If we could, in some way, alter the inertial mass of a body, we should apparently be justified in supposing that we would thereby alter its gravitational mass to just the same extent, so strictly and invariably connected are they.

This fact suggests the apparently wild idea that in talking of the inertial and gravitational masses of a body we are really talking of the same thing, as if we spoke of a body's "size" and also of its "magnitude." We say "apparently wild idea" because, if it were true, we should have to say that a body's gravitation was merely a manifestation of its inertia. The idea is so wild that it seems to have occurred to only one man in the history of science. As a result we have the tremendous scientific revolution called Einstein's General Principle of Relativity. At the present stage of our discussion, however, we shall ignore this development, and assume that we have satisfactorily isolated a fundamental property of

31

matter called mass, referring by this term indifferently either to inertial or gravitational mass, as we are justified in doing because of their absolutely strict proportionality.

Having isolated the notion of mass, Newton proceeded to give a complete account of the motions of material bodies conceived as masses moving in space and time. We must now therefore pay some attention to his conceptions of space and time. It is best to give these conceptions in his own words:

" Absolute, true, and mathematical time, of itself, and from its own nature, flows equably without regard to anything external, and by another name is called duration: relative, apparent, and common time is some sensible and external (whether accurate or unequable) measure of duration by the means of motion, which is commonly used instead of true time; such as an hour, a day, a month, a year.

" Absolute space, in its own nature, without regard to anything external, remains always similar and immovable. Relative space is some movable dimension or measure of the absolute spaces which our senses determine by its position to bodies, and which is vulgarly taken for immovable space; such is the dimension of a subterraneous, an aerial, or celestial space, determined by its position in respect of the earth. Absolute and relative space are the same in figure and magnitude; but they do not remain always numerically the same. For if the earth, for instance, moves, a space of our air, which relatively and in respect of the earth always remains the same, will at one time be one part of the absolute space into which the air passes; at another time it will be another part of the same, and so, absolutely understood, it will be perpetually mutable.

" Place is a part of space which a body takes up

and is, according to the space, either absolute or relative. . . .

"Absolute motion is the translation of a body from one place into another; and relative motion, the translation from one relative place into another. Thus in a ship under sail, the relative place of a body is that part of the ship which the body possesses; or that part of its cavity which the body fills, and which therefore moves together with the ship: and relative rest is the continuance of the body in the same part of the ship, or its cavity. But real, absolute rest is the continuance of the body in the same part of that immovable space in which the ship itself, its cavity and all that it contains, is moved. Wherefore, if the earth is really at rest, the body, which relatively rests in the ship, will really and absolutely move with the same velocity which the ship has on the earth. But if the earth also moves, the true and absolute motion of the body will arise, partly from the true motion of the earth in immovable space; partly from the relative motion of the ship on the earth; and if the body moves also relatively in the ship, its true motion will arise, partly from the true motion of the earth in immovable space, and partly from the relative motions as well of the ship on the earth as of the body on the ship; and from these relative motions will arise the relative motion of the body on the earth. . . .

"Absolute time, in astronomy, is distinguished from relative, by the equation or correction of the vulgar time. For the natural days are truly unequal, though they are commonly considered as equal, and used for a measure of time; astronomers correct this inequality for their more accurate deducing of the celestial motions. It may be that there is no such thing as an equable motion, whereby time may be accurately measured. All motions may be accelerated and retarded, but the true or equable progress of absolute time is liable to no change. The duration or perseverance of the existence of things remains the same; whether the motions are swift or slow, or none at all: and therefore it ought to be distinguished from what are only sensible measures thereof; and out of which we collect it, by means of the astronomical equation. The necessity of which equation, for determining the times of a phenomenon, is evinced as well from the experiments of the pendulum clock, as by eclipses of the satellites of Jupiter.

" As the order of the parts of time is immutable, so also is the order of the parts of space. Suppose those parts to be moved out of their places, and they will be moved (if the expression may be allowed) out of themselves. For times and spaces are, as it were, the places as well of themselves as of all other things. All things are placed in time as to order of succession; and in space as to order of situation. It is from their essence or nature that they are places; and that the primary places of things should be movable is absurd. These are, therefore, the absolute places, and translations out of those places are the only absolute motions.

" But because the parts of space cannot be seen, or distinguished from one another by our senses, therefore in their stead we use sensible measures of them. For from the positions and distances of things from any body considered as immovable, we define all places: and then with respect to such places, we estimate all motions, considering bodies as transferred from some of those places into others. And so instead of absolute places and motions we use relative ones; and that without any inconvenience in common affairs: but in philosophical disquisitions, we ought to abstract from our senses and consider things themselves, distinct from what are only sensible measures of them. For it may be that there is

no body really at rest, to which the places and motions
of others may be referred."

It seems strange that Newton should postulate the
existence of two entities, absolute time and absolute space,
which cannot, as he confesses, be observed. Newton
was an empiricist. He did not in the least believe that
laws of Nature could be deduced from *a priori* considerations. We have to begin, in every case, with observation
and experiment. How, then, did Newton come to assume
the existence of an unobservable absolute time and absolute space? It was because Newton thought that he had
observed absolute motion, and absolute motion implies
absolute time and absolute space. We will quote his
description of the celebrated experiment with a pail of
water:

" The effects which distinguish absolute from relative
motion are the forces receding from the axis of circular
motion. For there are no such forces in a circular
motion purely relative, but in a true and absolute circular
motion they are greater or less, according to the quantity
of the motion. If a vessel, hung by a long cord, is so
often turned about that the cord is strongly twisted, then
filled with water, and held at rest together with the
water; after, by the sudden action of another force, it is
whirled about the contrary way, and while the cord is un-
twisting itself, the vessel continues for some time in this
motion; the surface of the water will at first be plane,
as before the vessel began to move; but the vessel, by
gradually communicating its motion to the water, will
make it begin sensibly to revolve, and recede by little
and little from the middle, and ascend to the sides of the
vessel, forming itself into a concave figure (as I have
experienced), and the swifter the motion becomes, the
higher will the water rise, till at last, performing its revolutions in the same times with the vessel, it becomes

relatively at rest in it. This ascent of the water shows its endeavour to recede from the axis of its motion; and the true and absolute circular motion of the water, which is here directly contrary to the relative, discovers itself, and may be measured by this endeavour. At first, when the relative motion of the water in the vessel was greatest, it produced no endeavour to recede from the axis: the water showed no tendency to the circumference, nor any ascent towards the sides of the vessel, but remained a plane surface, and therefore its true circular motion had not yet begun. But afterwards, when the relative motion of the water had decreased, the ascent thereof towards the sides of the vessel proved its endeavour to recede from the axis; and this endeavour showed the real circular motion of the water perpetually increasing, till it had acquired its greatest quantity when the water rested relatively in its vessel. . . ."

This experiment, it is certain, cannot be interpreted as a case of relative motion between the pail and the water. For at first, when the pail begins to spin, and before the water has taken up the motion, there is relative motion between the pail and the water, and the surface of the water is flat. In the last stage, after the water has taken up the motion of the pail and become depressed in the middle, the pail is suddenly stopped. The water continues to rotate, so again there is relative motion between the pail and the water. But this time the surface of the water is not flat, but depressed in the middle. Between these two cases, therefore, we have an observable physical difference. Yet, if the motion between pail and water is to be regarded as purely relative there is no difference between the two cases. Newton therefore concluded, logically enough, that the motion was absolute. It might be objected to his conclusion that, after all, the pail of water is rotating with respect to the fixed stars.

This objection has some validity if it be meant to assert that rotation, like translation, is motion with respect to some point of reference. But it can hardly be interpreted to mean that the rotation of the pail of water is relative to the fixed stars in the sense that we could equally well assume the pail of water to be at rest and the whole stellar universe rotating round it. We may say that, as a matter of fact, we have never observed a case of rotation " in empty space." Newton imagined that if his experiment were performed in a universe containing nothing but the pail of water the same result would be obtained. We cannot make such an assumption. The experiment is performed in a universe containing, amongst other things, the fixed stars. We cannot say what would happen in a universe not containing those bodies. Newton, although he based his dynamical laws on observation, assumed that they held good in all circumstances. Thus that property of a body called its inertia, in virtue of which it persists in its state of rest or of uniform motion in a straight line, unless acted on by some force, is assumed by Newton to be an absolute property of the body, in the sense that it would be just the same in empty space at an infinite distance from all other bodies. Such assumptions are very natural, but one important aspect of Einstein's work is that he insists that none but observable factors shall be invoked in explaining phenomena, and that we shall not dogmatize about what would happen in conditions we can never experience. Thus Einstein does not regard Newton's absolute space as a legitimate entity to import into scientific descriptions, since it cannot be observed, and he refuses to base arguments on the assertion that a body's inertia would remain unaltered in certain conditions, when those conditions are forever inaccessible to observation.

In working out his system of dynamics and in applying

it to astronomy Newton makes no use of his notion of absolute space, apart from this question of rotation. His laws do not require the notion of absolute position. And his notion of absolute time is only used as a sort of inaccessible ideal towards which time-keeping instruments approximate. Consider, for instance, his first law of motion. It states that a body acted on by no forces remains at rest or in a state of uniform motion in a straight line. The word " rest " here does not mean rest in absolute space. The system of reference in which the body is at rest may itself be in motion. The notion of " uniform " motion is the notion that equal distances are traversed in equal times. This notion of " equal " times doubtless makes reference to Newton's absolute time, whose intervals are measured only approximately even by the most accurate of clocks. Newton doubtless thought his law to be rigorously true in the sense that it would be verified by yard measures and clocks made by God Himself. And it would be verified more and more accurately as man made or discovered more and more accurate measuring instruments. But a more important aspect of this law, from the modern point of view, is that it makes reference to an unobservable condition of affairs. It speaks of a body " acted on by no forces." No such body can be observed. Every body of which we have experience exists in the neighbourhood of gravitating masses. Newton, of course, explained the fact that we do not observe bodies moving in straight lines with uniform velocities by saying that they were acted on by forces. What Newton has really done, in enunciating his first law of motion, is again to make appeal to empty space. He states that in certain unobservable conditions, namely, at an infinite distance from all gravitating masses, a body, thrown into space, would move in a straight line with uniform velocity. Having assumed this he then explains the fact that observed bodies, such as planets, and stones thrown into the air, do not move in this way, by inventing a force of gravitation that pulls them out of their straight line course. Thus we may say that Newton's force of gravitation is made inevitable by assuming his first law of motion and then trying to make this law relevant to the observed behaviour of bodies.

Theoretically speaking, an entirely different approach to the whole question could have been made, as Einstein has shown. But it is highly improbable that Einstein or anybody else, in Newton's place and time, could have done other than adopt his conceptions. Einstein's outlook, in the absence of the necessary mathematical technique, could have been no more than a vague and disturbing intuition, impossible to formulate intellectually.

Newton continued on the path opened up by Galileo. He succeeded in isolating those of our perceptions between which quantitative relations exist, and in accounting for all the main phenomena of motion in terms of them. As we have seen, his procedure necessitated the invention of the force of gravitation. In terms of these concepts he accounted for all the chief phenomena of dynamical astronomy. Kepler's three laws of planetary motion, for instance, the results of an interminable trial and error process, were shown to be necessary consequences of Newton's fundamental principles. It may be mentioned, incidentally, that Newton gave geometrical solutions of all his chief problems. But the detailed investigation of motion necessitated the invention of a more flexible mathematical instrument than was provided by classical geometry. The notions of " continuity " and of " rate of change " are so fundamental in physics that their correct formulation may be regarded as being as important as the formulation of the notion of mass. Attempts to formulate them led to the invention of the differential calculus, one of the most important of all scientific instruments. In this formulation Newton played a leading rôle. With this instrument and with his definitions and laws, Newton bequeathed to posterity a method of unequalled generality for the mathematical description of natural phenomena.

A cardinal feature of this whole achievement must now be made explicit. Although Newton denied " action at a distance " his equations assume it. The gravitational force acting on a body and due to surrounding bodies is, at any given instant, determined by the configuration of those bodies at that same instant. For the prediction of future phenomena we require to know nothing beyond the instantaneous distribution of masses throughout space at a given instant. From this configuration of masses future configurations can be deduced. Thus a certain conception of space and time is assumed. There is one universal space and there is one universal time. Further, the relation of a material body to space and time is the relation that Dr. Whitehead [1] calls " simple location." We can say of a material body that it is " here " in space and " here " in time without making reference to other regions of space and time. That is to say, the body is wholly itself at any instant. It simply persists in time, for instance, but it owes nothing of its essential nature to time. The whole of it exists at any period of time, however short. In this respect, therefore, a piece of matter is quite unlike a tune, which requires a certain period of time to exist at all. Whether the Newtonian view of matter is the true one, or whether we have to introduce into our conception of matter some of the characteristics of a tune,

[1] A. N. Whitehead: *Science and the Modern World.*

is a question to be discussed later. At present the point is raised merely to suggest that the Newtonian conceptions of space, time, and matter, natural as they seem and successful as they have been, are not without alternatives.

CHAPTER III

THE ETHER THEORY

THE attempt to describe Nature mathematically had, as a result of Newton's efforts, attained a great measure of success so far as the phenomena of motion were concerned. All the main motions of the celestial bodies had been given a satisfactory mathematical description, and many of the motions that occur on the surface of the earth had been equally well described. But there remained a vast range of phenomena for which a mathematical description had not yet been given. There were the phenomena of light and heat and of the electric and magnetic attractions. Other things that puzzled Newton were " why two well-polished metals cohere in a receiver exhausted of air; why mercury stands sometimes up to the top of a glass pipe, though much higher than thirty inches. . . . Why the parts of all bodies cohere; also the cause of filtration, and of the rising of water in small glass pipes above the surface of the stagnating water they are dipped into." He also wanted to know how " all sensation is excited," and how " the members of animal bodies move at the command of the will." For all these phenomena Newton had recourse, provisionally at any rate, to the notion of an " Ether," a notion that was destined to become one of the major scientific concepts. Newton's conception of the Ether underwent various modifications, but it never advanced sufficiently far to receive mathematical formulation. He outlines an early idea of it as follows:

" It is to be supposed that there is an ethereal medium,

42

much of the same constitution with air, but far rarer, subtler, and more strongly elastic. But it is not to be supposed that this medium is one uniform matter, but composed partly of the main phlegmatic body of ether, partly of other various ethereal spirits, much after the manner that air is compounded of the phlegmatic body of air intermixed with various vapours and exhalations. For the electric and magnetic effluvia, and the gravitating principle, seem to argue such variety. Perhaps the whole frame of Nature may be nothing but various contextures of some certain ethereal spirits or vapours, condensed as it were by precipitation, much after the manner that vapours are condensed into water, or exhalations into grosser substances, though not so easily condensable; and after condensation wrought into various forms, at first by the immediate hand of the Creator, and ever since by the power of Nature, which, by virtue of the command, increase and multiply, became a complete imitator of the copy set her by the Protoplast. Thus perhaps may all things be originated from ether."

This last sentence is particularly significant, for by reducing all matter to ethereal " condensations," and re-garding the ether as constituted by minute particles, New-ton had the idea that possibly all phenomena could ultimately be explained in terms of attractions and re-pulsions between these particles, and thus a complete mathematical description of the world be given. Besides requiring the ether to explain the phenomena described above, Newton required it to serve two other purposes. He required the ether to account for gravitation and also for the fact that the universe is not at a standstill. On this latter point Newton remarks that inertia, or *vis inertiae*, as he calls it, " is a passive principle by which bodies persist in their motion or rest, receive motion in proportion to the force impressing it, and resist as much as they are resisted." But, he goes on:

"By this principle alone there never could have been any motion in the world. Some other principle was necessary for putting bodies into motion; and now they are in motion, some other principle is necessary for conserving motion. For from the various composition of two motions, 'tis very certain that there is not always the same quantity of motion in the world. . . . But by reason of the tenacity of fluids, and attrition of their parts, and the weakness of elasticity in solids, motion is much more apt to be lost than got, and is always upon the decay. For bodies which are either absolutely hard, or so soft as to be void of elasticity, will not rebound from one another. Impenetrability makes them only stop. If two equal bodies meet directly in vacuo they will by the laws of motion stop where they meet, and lose all their motion, and remain in rest, unless they be elastic, and receive new motion from their spring. If they have so much elasticity as suffices to make them rebound with a quarter, or half, or three quarters of the force with which they come together, they will lose three quarters, or half, or a quarter of their motion."

And not only such cases, but the heat of the sun, the beating of the heart, etc., are all expenditures of energy which require to be kept up. They are kept up by what Newton calls "active principles."

"Seeing therefore the variety of motion which we find in the world is always decreasing, there is a necessity of conserving and recruiting it by active principles, such as are the cause of gravity, by which planets and comets keep their motions in their orbs, and bodies acquire great motion in falling; and the cause of fermentation, by which the heart and blood of animals are kept in perpetual motion and heat; the inward parts of the earth are constantly warmed, and in some places grow very hot; bodies burn and shine, mountains take fire, the caverns of the earth are blown up, and the sun continues violently hot and lucid, and warms all things by his light. For we meet with very little motion in the world, besides what is owing to these active principles. And if it were not for these principles the bodies of the earth, planets, comets, sun, and all things in them would grow cold and freeze, and become inactive masses; and all putrefaction, generation, vegetation, and life would cease, and the planets and comets would not remain in their orbs."

Newton knew nothing, of course, of the principle of the conservation of energy. Many of the remarks in the above passage are due to this fact. That the heart and blood of food-taking animals are kept in " perpetual motion and heat," for instance, is no longer a mystery. But the mere fact that energy is conserved does not meet all Newton's points. For it is not the total energy but the amount of available energy that his argument is concerned with. The principle of the conservation of energy does not solve the problem presented by the fact that " the sun continues violently hot and lucid." In the main, however, the passage is a testimony to Newton's ignorance of the conservation of energy. Instead of that principle, he invoked the ether as the seat of the requisite " active principles." The ether's other function, that of accounting for gravity, is accomplished as follows:

" For this end I will suppose ether to consist of parts differing from one another in *subtilty* by indefinite degrees; that in the pores of bodies there is less of the grosser ether, in proportion to the finer, than in the regions of the air; and that yet the grosser ether in the air affects the upper regions of the earth, and the finer ether in the earth the lower regions of the air in such a manner, that from the top of the air to the surface of the earth, and again from the surface of the earth to the centre thereof, the ether is insensibly finer and finer. Imagine now any body suspended in the air, or lying on the earth, and the ether being by the hypothesis grosser in the pores, which are in the upper parts of the body, than in those which are in its lower parts, and that grosser ether being less apt to be lodged in those pores than the finer ether below, it will endeavour to get out and give way to the finer ether below, which cannot be, without the bodies descending to make room above for it to go out into."

These speculations were not taken very seriously by Newton. He made them unofficially, as it were. He did not, in his official capacity, profess to give any " explanation " of gravitation at all. In fact, in some moods, he did not consider it to be the function of science to give such explanations. His equations described the observed phenomena and that was all that science could do. Newton, indeed, seems at times to have had strong leanings towards a point of view that was much later made explicit by certain continental mathematicians, the view that all we can ever know of Nature is the equations summarizing its observed behaviour. Of the *meaning* of these equations it is useless to enquire. But the majority of scientific men have not been able to adopt this outlook. They look for " explanations," that is, for descriptions couched in terms of the familiar properties of matter. From the purely logical point of view this seems to be a monstrous demand. Why should we expect certain properties of gross matter to furnish the key to all natural processes? Lord Kelvin stated that he could understand nothing of which he could not make a mechanical model. In this he was by no means alone. This conception of explanation underlay the development of the great ether theory. It is characteristic of modern physics that this criterion is being much less insisted on. Of the two Newtons, the one who austerely refused to pass beyond his equations, and the other who played with an ether "much of the same constitution with air," it is the first whose spirit dominates modern science.

The earliest conceptions of the ether, as we see even from Newton's remarks on it, were exceedingly vague. It was merely a kind of hold-all for exhalations and emanations of all kinds. Vortical whorls could be set up in it and also wave motions. The conception of it was decidedly material, but could hardly be called mechanical. Gilbert, for instance, the great father of electrical science, supposed electric attraction to be exerted by an emanation from the electrified body, the emanation having an inherent tendency to reunion with the body. Descartes explained magnetism by supposing a vortex of fluid matter to surround the magnet, the vortex entering by one pole and leaving at the other. Iron and steel offered a special resistance to the motion of this vortex and were therefore carried along by it. Hooke and Huygens considered that light was a wave motion in the ether. But none of these theories provided a mechanism for the ether that should explain the observed results. Newton himself developed a corpuscular theory of light, rejecting the wave theory for reasons that were perfectly valid at his time. But all these theories were, for Newton, merely ways of questioning Nature. He held none of them with any great tenacity. All of them were, in truth, too vague. The state of the ether theory of light at that time, and Newton's attitude towards it, may be illustrated by the following quotation. Speaking of light, he says:

"They, that will, may suppose it an aggregate of various peripatetic qualities. Others may suppose it multitudes of unimaginable small and swift corpuscles of various sizes, springing from shining bodies at great distances one after another; but yet without any sensible interval of time, and continually urged forward by a principle of motion, which in the beginning accelerates them, till the resistance of the aethereal medium equals the force of that principle, much after the manner that bodies let fall in water are accelerated till the resistance of the water equals the force of gravity. But they, that like not this, may suppose light any other corporeal emanation, or any impulse or motion of any other medium or aethereal spirit diffused through the main body of ether, or what else they can imagine proper for this purpose."

The contrast between this kind of vague fumbling and the crystalline clarity of the Newtonian dynamics is remarkable. So far as the phenomena of gross matter in motion are concerned, science is fully mature. The mathematical description of the rest of phenomena seems hardly to have been begun. The next great achievement of the scientific adventure was to introduce mathematical order into the phenomena of light. This great achievement was carried through without introducing any radically new concepts into the scientific scheme. The Newtonian set of abstractions, mass, force, etc., were found to be sufficient. But this sufficiency was only secured by postulating an ether and giving this ether definite properties. As these properties were investigated they became more and more remarkable until the ether became finally a wholly unnatural monster. The ether may be regarded as a vast by-product of the attempt to describe Nature mathematically. Scientific men were saddled with this frightful creation in order that the phenomena of light should lend themselves to mathematical description in terms of the Newtonian set of abstractions. This vast and complicated and, finally, incredible entity was the incidental product of man's desire for simplicity or,

48

rather, of his desire that explanation should proceed in terms of the familiar.

The question of the mechanism of the propagation of light was made acute when the Danish astronomer Römer found, in 1676, that this propagation occurs with a finite velocity. He was led to assume this by observing the eclipses of the satellites of Jupiter. As these satellites pass behind the body of Jupiter they vanish from our view. The interval between successive disappearances can be calculated. Römer noticed that, when the earth was moving away from Jupiter, the eclipse interval was increased. When the earth was approaching Jupiter the eclipse interval was diminished. If light is not transmitted from Jupiter's satellites to the earth instantaneously these differences can be accounted for. In the first case the light has to overtake the retreating earth; in the second case the earth is advancing to meet the light. From his observations Römer calculated that light travelled with a velocity of 192,000 miles per second. It was objected to Römer's deduction that his observations might be due to inequalities in the motions of the satellites. Later on, however, experiments carried out on the surface of the earth put the finite propagation of light beyond all doubt. This discovery, as we have said, makes the mechanism of light's propagation a matter of serious interest. Light takes eight minutes to arrive at the earth from the sun. It is natural to suppose that it occupies that time in traversing the intervening space. In what form does it exist in that space? How is it transmitted?

Two theories, both of great antiquity, have been proposed to account for the transmission of light. One is the corpuscular theory; the other is the wave theory. The corpuscular theory supposes light to consist of tiny particles shot out in all directions by the luminous body. The other theory supposes that light consists of waves

49

in an ether, a medium filling all space and penetrating all bodies. Newton paid much attention to the corpuscular theory, and Huygens is usually regarded as the first convincing exponent of the wave theory. The triumph of the wave theory was long hindered by the assumption that the ether was a sort of air, able to support only longitudinal waves, such as occur when sound is propagated. On this hypothesis many optical phenomena remained unexplained. It was found, however, that these recalcitrant phenomena could be accounted for if light be supposed to consist of transverse rather than of longitudinal waves, that is, of waves whose vibrations are at right angles to the direction of propagation (as occurs when one end of a rope be shaken, the other end being fixed) and not, as with longitudinal waves, in the line of propagation. And from this it immediately became apparent that the ether must be conceived, not as a sort of air, but as a sort of jelly.

Thus arose the great period of ether theory, the theory of the ether as an " elastic solid." With this development the theory of light passed definitely into the hands of the mathematicians, and something like a mathematical orgy broke out. Many of the finest mathematicians of the nineteenth century spent immense pains and ingenuity in working out the mechanical properties of the ether. They began by assigning to the ether properties that may be called probable, in the sense that they were not unlike the properties of ordinary matter. But such simple ethers were not competent to explain all the known optical phenomena. More and more extraordinary properties had to be given to the ether; it became more and more unlike anything of which scientific men had experience. Sometimes the analogy to ordinary experience was frankly given up. Sometimes quite contradictory physical properties were attributed to the ether. Mathematicians would develop their etheric theories very much for " the love of it," working out the consequences of their equations with great ingenuity and singleness of mind, leaving in their rear, like unconquered fortresses, fundamental and unsolved physical difficulties. For some time, for instance, nobody met the difficulty that confronts the elastic solid theory of the ether at the outset. That difficulty is to explain how, if the ether has the properties of a solid, the planets are able to journey through it at immense speeds without encountering any perceptible resistance. Sir George Stokes found the answer in cobbler's wax. Here is a substance rigid enough to be capable of elastic vibration and yet plastic enough to permit bodies to pass slowly through it. By imagining the ether to have, to an immensely exaggerated extent, the properties of cobbler's wax, the difficulty aroused by the motions of the planets could be resolved.

But other difficulties remained. Does the ether within material bodies possess the same properties as the ether outside? Is the ether in the neighbourhood of moving matter carried along by the matter? Some observations seemed to show that it was, others that it was not. On the whole, the theory of the ether represents one of the most laborious and least satisfactory expenditures of ingenuity in the history of science. No ether was constructed by the mathematicians that fulfilled all the demands made on it. As the ether became more complicated it became more unplausible and unaesthetic. Space became filled with an incredible combination of cogwheels, gyroscopes and driving bands. It resembled nothing so much as the nightmare of some mad engineer. It is doubtful whether anything of real importance has emerged from the whole of this nineteenth-century ether except the experiments that uniformly failed to detect its existence. This particular aspect of the great effort to describe Nature mathematically must be considered, on the whole, a failure. With the rise of Maxwell's electromagnetic theory of light the ether theory acquired a subordinate position, and with the modern theory of relativity it has become a mere ghost. Accompanying the decline of the ether has occurred a change in the scientific attitude towards " explanations." The mechanical properties of matter no longer hold the fundamental position they held in the science of the nineteenth century, and it is not considered necessary that descriptions of all the fundamental processes of Nature should be in terms of them. Indeed, these same properties of matter are now regarded as " derived " concepts. Quite different concepts form the basis of modern physical theories.

By the time that the elastic solid theory of the ether began to attract the attention of mathematicians the Newtonian scheme for explaining the phenomena of motion had received an immense development. In the hands of Lagrange, Newton's laws of motion received a more general and more compact form of statement. The actual phenomena of motion, as exhibited in the solar system, for instance, were, owing to the efforts of many mathematicians, of whom Lagrange and Laplace were the chief, worked out in many cases of great complexity. It became more and more apparent that the Newtonian dynamical scheme provided the true key for the understanding of all the phenomena of motion. Dynamical astronomy was merely the most grandiose example of the power of the great theory of attractions. Two of the fundamental concepts of this theory were those of " particles " and " attracting forces." A particle was conceived as a tiny piece of matter, possessing mass, but so small that it could be treated in calculation as a mathematical point. Material bodies were conceived as made up of these particles and the total effect of a body at an external point was conceived as the sum of the effects exerted by its constituent particles. Thus one of Newton's most beautiful theorems was his proof that the gravitational attraction of a sphere at an external point was the same as if the whole mass of the sphere was concentrated at its centre. This method of analysis, in terms of particles and attracting forces, was applied to a variety of problems. Thus Laplace tried to account for the phenomena of capillarity in these terms, but a much more significant exhibition of the power of these concepts is to be found in Poisson's application of them to the phenomena of electricity and magnetism. It was known that electrified bodies sometimes attracted one another and sometimes repelled one another, and these effects were supposed to show that two kinds of electric " fluids " existed. There was also a one-fluid theory that could be made to explain the facts, and it was impossible to decide, experimentally, which theory was right. The law of attraction and repulsion between electrified bodies was known. The way in which the repulsive and attractive forces between electrified bodies varied with the distance between them had been determined experimentally by Coulomb, and he had found the law of variation to be the same as that for gravitation, namely, inversely as the square of the distance. The immense mathematical apparatus that had been built up to deal with the Newtonian theory of gravitation was therefore, with certain modifications, at the disposal of the theory of statical electricity. Poisson adopted the two-fluid theory of electricity, a perfectly good basis for his mathematical deductions, since they remain valid even if the two-fluid theory be abandoned. The fluids existed merely to provide some sort of tangible agency for the attractive and repulsive forces. Poisson opens his memoir by saying:

" The theory of electricity which is most generally accepted is that which attributes the phenomena to two different fluids which are contained in all material bodies. It is supposed that molecules of the same fluid repel each other and attract the molecules of the other fluid; these forces of attraction and repulsion obey the law of the inverse square of the distance; and at the same distance the attractive power is equal to the repellent power; whence it follows that, when all the parts of a body contain equal quantities of the two fluids, the latter do not exert any influence on the fluids contained in neighbouring bodies, and consequently no electrical effects are discernible. This equal and uniform distribution of the two fluids is called the *natural state*; when this state is disturbed in any body, the body is said to be *electrified*, and the various phenomena of electricity begin to take place."

Poisson's paper is concerned wholly with electrostatics, a science which deals with a very small region of natural phenomena, but a region in which the mathematician can go on from one ingenuity to another. Poisson wrote an equally masterly paper on magnetism, dealing with another restricted region of almost equally ideal simplicity. We say "ideal simplicity," for such investigations, complicated as they may be mathematically, are remote from the much more complicated conditions that actually occur in Nature. These spheres, spheroids, and ellipsoids, ideal geometric figures, whose properties are investigated, are merely first, and often not very close, approximations to the conditions that actually occur in nature and in technical applications. But although a particular investigation may sometimes appear more of a mathematician's pastime than an inquiry into Nature, it is often found to have an unsuspected value. The development of electrostatics, unreal as some of the problems were, led to mathematical theorems of great beauty and generality which were found applicable to a great variety of phenomena.

One point in these investigations that deserves to be

noticed is that, like the theory of gravitation itself, they all assumed "action at a distance." The attracting and repelling bodies, whether particles or fluids, were all assumed, in the formulae, to influence one another instantaneously. Similar assumptions were made by Ampère, in the celebrated memoir in which he reduced the phenomena of *electric currents* to mathematics. This was a very great mathematical triumph. Maxwell, an austere judge, referred to it, half a century afterwards, as "one of the most brilliant achievements in science," and goes on: "The whole, theory and experiment, seems as if it had leaped, full-grown and full-armed, from the brain of the 'Newton of electricity.' It is perfect in form and unassailable in accuracy; and it is summed up in a formula from which all the phenomena may be deduced, and which must always remain the cardinal formula of electrodynamics." From our present point of view it is interesting to note that Ampère assumes, throughout this memoir, action at a distance. He proclaims himself a follower of that school that endeavours to explain all physical phenomena in terms of equal and oppositely directed forces between pairs of particles. But he is quite aware of the fact that a more fundamental explanation may have to be sought for in different terms. Thus he thinks that these electric phenomena may be due to "the reaction of the elastic fluid which extends throughout all space, whose vibrations produce the phenomena of light." But this is merely a speculation; for the purposes of calculation he assumes action at a distance.

We must remember that a great deal of this early work was, in a sense, formal. Phenomena were to be treated mathematically. For this purpose certain simplifying assumptions were made. These assumptions were not supposed so much to be assertions of actual fact, as to be in *formal* correspondence with the facts. Thus it was not necessarily supposed that electricity actually consisted of two " fluids," but it was asserted that observed electrical phenomena were compatible with mathematical deductions from that hypothesis. A very different hypothesis might lead to the same mathematical deductions, and therefore be quite as consistent with observation. But, in order to get on with the mathematical description as swiftly as possible, it was advisable to adopt the simplest hypothesis which could serve as a sufficient basis. Similarly, the Newtonian force of gravitation was accepted, for mathematical purposes, as an ultimate fact, but it was not thereby intended to imply that action at a distance was an ultimate property of matter. In these departments of science the mathematical interest swamped, as it were, the physical interest. The luminiferous ether, on the other hand, was intended, on the whole, to represent in intimate detail the actual mechanism of the propagation of light.

Thus we see that by the beginning of the nineteenth century the sciences of motion, of light, and of electromagnetism existed at different levels, as it were, of completeness. The phenomena of matter in motion were, it was generally believed, completely understood. The properties of mass, extension, impenetrability, and so on, attributed to matter, were regarded as fundamental characteristics of an enduring substance. Whatever other properties matter might possess, it certainly possessed these. And these characteristics, together with the notion of force, entered into any complete mathematical description of matter in motion. So far as the universe consists of masses in motion, therefore, it could be completely understood. Order and coherence had thereby been bestowed upon a wide range of phenomena. Mystery had been abolished throughout a wide region, from the movements of planets and comets to the behaviour of rotating flywheels and colliding billiard balls. But for this to be possible certain subsidiary characteristics of matter had to be introduced. The most fundamental characteristic of matter involved in these descriptions was "inertia." This was regarded as a fundamental and absolute property of matter. Other characteristics of matter were involved in particular cases of motion. The particles of the rotating flywheel, for instance, possessed the property of "cohesion." The colliding billiard balls possessed the property of "elasticity." The planets and comets possessed the property of "gravitation." It was considered that all these properties were possibly manifestations of more fundamental properties. A knowledge of the intimate structure of matter might explain cohesion and elasticity in other terms. A complete knowledge of the properties of the ether might show that gravitation need not be regarded as an ultimate property of matter. Inertia, however, was regarded as ultimate. It was so indissolubly linked up with the notion of matter that it could be used to define matter. Matter could be defined as that which possesses inertia. When, therefore, men asked whether light, heat, and the electric fluids were "material" the question could at once be answered by finding out whether they possessed inertia.

The phenomena of light were not as well understood as the phenomena of motion. That light consisted of transverse waves in an elastic solid was generally admitted, but the properties of this elastic solid remained somewhat elusive. No perfectly clear and comprehensive description of it could be given. The hypothesis had achieved such great triumphs that it was generally believed to correspond closely to the actual workings of Nature. But these actual workings seemed to be extremely complicated. It became a little difficult to believe that Nature was really as complicated as all that. One became suspicious that these difficulties must be due to a wrong method of approach. On the other hand, one had to admit the possible truth of Fresnel's dictum that Nature cares nothing for analytical difficulties. The human ideal of simplicity may be quite irrelevant where the works of God are concerned. It is very relevant, however, to the construction of a science. The monstrous complications of the ether theory afford no rational ground for objecting to it. But they have the very important effect of making the theory unaesthetic and therefore uninteresting. If all scientific descriptions of phenomena proceeded in terms of such entities as the elastic solid ether it is doubtful whether many men would take up science at all. The ether theory began to be neglected before all its possibilities were exhausted, and largely, we may suppose, because it began to be boring. As an explanation of the phenomena of light, therefore, the ether theory, at the time we are discussing, was a sort of half-way house. It was an attempt to make the fundamental material concepts, isolated by Newton, extend to this different department of phenomena. It was, at bottom, an attempt to discover whether the concepts found sufficient to describe the phenomena of motion were sufficient also to describe the phenomena of light.

The explanation of electromagnetic phenomena was of a different order. Here none but the formal characteristics of the phenomena were described. The theory that electricity consisted of two fluids, or of one fluid, was not regarded as an exhaustive account of the physical nature of electricity. Electricity was likened to a fluid merely because it was mobile. The essential feature of the hypothesis was that each little area of an electrified surface attracted or repelled any other little electrified area with a force that varied inversely as the square of the distance. Nothing was really asserted about the constitution of electricity or about the manner in which the force was exerted. The mathematical skeleton was all that interested such a man as Ampère. The same mathematical skeleton could support any number of physical bodies. Action at a distance was assumed, merely because the assumption was mathematically simple and no phenomena were known that made it inadmissible. The theory was very much less penetrating and exhaustive than the wave theory of light. In fact, it can hardly be called a theory at all. It was rather an exposition of the mathematical framework within which any physical theory must be constructed. In this respect the theory would be considered by some people to belong to the first, and by others to the last, stage of science. To a scientific man like Lord Kelvin such a theory is merely a clearing of the ground. He would immediately proceed to imagine mechanical models, and probably even to construct them in a workshop, that should enable him to picture, in terms of familiar images, the phenomena in question. To certain continental mathematicians, on the other hand, the mathematical framework is all we can ever know about Nature. Mechanical models may be made that obey this framework, but they tell us nothing we did not already know. Such properties as they possess independent of the framework need not correspond to anything in Nature. If Ampère's theory accounted for all electric phenomena its extreme abstraction would probably be no objection to it. Its imperfection lies in the fact that it does not account for all the known facts.

So far, as we see, no radically new conceptions have been introduced. The familiar properties of matter, mass, inertia, cohesion, elasticity, etc., together with the notions of attractive and repulsive forces, were adequate to explain the phenomena of motion, of light considered as waves in an elastic solid, and of electricity conceived as attracting and repelling fluids. It is really amazing the variety of phenomena to which the few and simple fundamental

Newtonian conceptions provided the key. We do not meet with a new conception in science until we come to consider the notion of " energy."

Although, as we have indicated, the hypothesis of an elastic solid ether has not turned out to be very satisfactory, it will be convenient to give, here, a brief account of the chief phenomena that led men of science to conclude that light is a wave-motion. The first of these phenomena is " Interference." As early as 1665 Grimaldi had stated that, in certain circumstances, two lights, when superposed, could destroy one another; but his experiment, as he describes it, does not warrant this conclusion. Dr. Young appears to have been the first to obtain a genuine example of interference. He caused a beam of light to pass through a narrow slit in a shutter. This beam then fell upon a screen perforated by two pin-holes very close together. The two small pencils of light that emerged from these pin-holes were then allowed to fall upon another screen. Young then observed that the place on the screen where these two pencils overlap is not evenly illuminated, but is covered by a series of bright and dark bands alternating with one another. This phenomenon can immediately be explained if we suppose each pencil of light to consist of a train of waves. For if two waves are superposed they may either reinforce or destroy one another. Where the trough of one wave coincides with the crest of the other we have a dark band; where the crests coincide we have a bright band. The wave theory explains this phenomenon of interference both qualitatively and quantitatively. No actual destruction of light is involved. The total energy of the two trains of waves is merely redistributed.

Another phenomenon which involves the wave theory of light is diffraction. It had been objected to the wave theory that objects cast sharp shadows. Light, like sound, should bend round obstacles. But this objection can be met if we suppose the waves of light to be extremely minute. It can then be shown that the degree of bending they would experience on passing a sharp edge would be very small. Such small deviations, in entire agreement with calculation, have been observed. This is the phenomenon of diffraction. What was intended to be an objection to the wave-theory, therefore, turns out to be a strong support of it.

That light waves are transverse, and not longitudinal, may be deduced from the phenomenon called polarization. Transverse vibrations are executed at right angles to the line of propagation. Thus a light ray may be conceived as a vibrating cord, except that the vibrations are occurring in all directions at right angles to the length of the cord. If such a cord were made to pass through a long narrow slot in a plate we see that only those vibrations that took place along the length of the slot could continue. Owing to the narrowness of the slot there would be no room for the vibrations in other directions. A crystal silicate called tourmaline seems to act upon light vibrations in just this way. Light which has passed through a plate of tourmaline is executing all its vibrations in one direction. Such light is said to be polarized. This polarized light may be passed through a second plate of tourmaline, but if this second plate be rotated round the ray of light as an axis the light gets fainter until, when the two plates are crossed at right angles to one another, no light gets through. It is as if we had our cord passing through two narrow slits at right angles to one another. Obviously no transverse vibrations could then get through at all. If light consisted of longitudinal vibrations, that is, of to and fro motions along the length of the cord, this effect would be inexplicable.

61

The hypothesis that light consists of transverse waves does represent, therefore, certain formal properties of light. It is the attempt to give this hypothesis a dynamical foundation in terms of the Newtonian concepts that has led to the unsatisfactory elastic solid theory of the ether.

CHAPTER IV

HEAT AND ENERGY

BOYLE, Hooke, Newton, and some other eminent men of the time held correct ideas on the nature of heat. They supposed the heat of a body to consist in the motion of its constituent particles. But this notion did not make headway against the eighteenth-century liking for "fluids." The theory of heat that reigned until the beginning of the nineteenth century was that heat was due to the action of an elastic and self-repellent fluid. In order to explain the observed facts the properties attributed to this fluid were a little complicated. Thus the particles of the fluid, besides repelling one another, were supposed to be attracted by particles of ordinary matter, but different kinds of matter attracted the heat particles differently. The expansion of bodies on being heated was explained as due to the self-repellent property of the heat particles introduced into the body. This did not, however, explain the fact that water expands on freezing. It was postulated, therefore, that the union of heat particles and material particles was of a chemical nature where the resulting change of volume is unpredictable. This had the satisfactory result of making it impossible to predict the consequent change of volume, and thus removed the theory, in this respect, from the region of test. Whether caloric, as this heat fluid was called, was or was not material, was a question that experiment was not at first sufficiently advanced to answer. Some experiments showed that a hot body weighed more than a cold one, some that a cold body weighed more than a hot one. It was not until the close of the eighteenth century that it was finally decided that the heat fluid was imponderable.

The vogue of the caloric theory, in spite of the absurd ingenuities to which its defenders were led, is one of the most curious episodes in the history of science. We see that the notion of " substance " had become almost an obsession. This notion was, historically, an inheritance from mediaeval thought. Psychologically it is derived chiefly from our tactile impressions. But the notion of substance was never made perfectly clear, although it was admitted that substances could be imponderable as well as ponderable, a distinction that cries aloud for a definition of substance. In the most modern scientific theories the notion has been largely, if not wholly, dispensed with.

The notion that heat was due to the presence of a substance was made extremely unlikely by the experiments performed by Count Rumford in 1798. He had noticed, while boring brass cannon at the military arsenal in Munich, that the metallic chips thrown off were very hot. He thereupon undertook experiments to find out how much heat could be developed by friction. He found that the amount of heat that could be obtained in this way was apparently inexhaustible. He says:

" In meditating over the results of all these experiments, we are naturally brought to the great question, which has so often been the subject of speculation among philosophers, namely, What is Heat?—is there any such thing as an *igneous* fluid? Is there anything that can with propriety be called *caloric*?

" In reasoning on this subject we must not forget that most remarkable circumstance, that the source of the heat generated by friction in these experiments appeared evidently to be *inexhaustible.*

" It is hardly necessary to add that anything which any *insulated* body or system of bodies can continue to furnish without *limitation* cannot possibly be a *material substance*; and it appears to me to be extremely difficult, if not quite impossible, to form any distinct idea of anything capable of being excited and communicated in the manner the heat was excited and communicated in these experiments, except it be *motion*."

Humphry Davy gave a further blow to the caloric theory by showing that two pieces of ice could be melted by rubbing them together. It is interesting to reflect that a sufficiently pertinacious and ingenious defender of the caloric theory could have met these objections by making his notion of caloric more abstract, and by doing so he would have stumbled on the modern notion of energy. For the property attributed to caloric, that it cannot be created or destroyed, is precisely what is asserted of energy, and the heat that appears in these experiments is not created out of nothing, but comes from the source that works the machinery. This, however, would not affect the fact that the form in which the energy manifests itself is as the motion of the constituent particles of the hot body.

The theory that the heat of a body is due to the motion of its constituent particles was quite concordant with the prevalent theory of the constitution of matter. The atomic theory of matter was given definite shape by Dalton very early in the nineteenth century. Matter, it appeared, was constituted of small indivisible particles called atoms. The indivisibility attributed to the atoms had reference, it must be emphasized, only to practical operations. No " metaphysical " indivisibility was asserted. The statement that atoms are the smallest particles of a substance merely means that the smallest portion of the substance that takes part in any known chemical reaction is called an atom. Strictly speaking, the term atom only applies to the smallest portions of " elementary " substances. The

c

smallest portions of " compound " substances, consisting, as they do, of two or more atoms united together, are called molecules. Thus one can have an atom of hydrogen, or an atom of chlorine, but not an atom of hydrogen chloride. A molecule contains two or more atoms. The atoms may be of the same kind, in which case we have a molecule of an elementary substance. All atoms of the same substance were supposed to be exactly alike. In particular, they had the same weight.

The atoms of the various elementary substances could be arranged in order of their weights, beginning with hydrogen, the lightest of them. When this was done it was found that a large proportion of the atomic weights were whole multiples of the weight of the hydrogen atom, or were very near such whole multiples. This fact supported the hypothesis, put forward by Prout, in 1815, that all the elementary substances were built up out of hydrogen. For the weight of the atom of any elementary substance would be, on this view, equal to the weight of the number of hydrogen atoms that composed it, and this number would be, of course, a whole number. More accurate determinations of atomic weights, however, showed so many departures from whole numbers that Prout's suggestion lost greatly in plausibility. Dumas tried to rehabilitate the theory by suggesting that half-atoms of hydrogen could be employed in building up other atoms. Even this would not square with the figures, so he suggested quarter-atoms of hydrogen. It is obvious that if this process be continued long enough it is bound to be successful, and that fact robs it of all plausibility. Nevertheless, Prout's theory, in a modified form, has been justified by modern researches on the atom, as we shall see later. The very departures from whole numbers, which ruined the simple harmony of Prout's scheme, have been shown to testify to the existence of much profounder harmonies. This is one more indication of the fact so frequently encountered in the history of scientific research, that what the theorist at the moment regards as recalcitrant and unaesthetic little details are, when properly understood, the key to a more beautiful world.

The atomic theory of matter, then, made heat as a mode of motion a perfectly plausible idea. The broad fact that increasing temperature changes the state of a body from the solid to the liquid, and then to the gaseous, fitted in perfectly with the new theory. For in a solid we may picture the constituent molecules as fairly strictly bound together. As the solid grows hotter its molecules move more violently, until finally they achieve the mobility of the liquid state. Still further agitation loosens the connections between the molecules still more, until they achieve the maximum freedom as a gas.

The theory of heat to which Rumford was led by his experiments, therefore, became the orthodox theory of science. But there is another aspect of his experiments which is of first-rate importance. Rumford's boring apparatus was driven by a horse. Is there any connection between the work done by the horse and the amount of heat developed in these experiments? The connection to be sought for is, of course, quantitative. If such a connection can be established the scientific adventure will have achieved another triumph, for, besides reducing heat to motion it will, in spite of this motion being chaotic and inaccessible to direct measurement, have succeeded in giving a mathematical description of heat. Can work be converted into heat? Can we say that the chaotic molecular motions of a certain body at a certain temperature can be precisely measured? The first great step in this investigation was made by Dr. Joule of Manchester in 1840. Leaden weights were made to descend through a known height and, by doing so, to turn a brass paddle revolving in water. The water was heated by the friction of the paddle, and the rise in temperature was measured by a delicate thermometer. Thus a known mass of water was raised through a known temperature by the performance of a known quantity of work. It could be assumed that the heat communicated to the water was equivalent to the work expended. It might be, of course, that not all the work goes to produce heat. It might be that slight electric and magnetic effects arise from the friction of the apparatus, and these effects might vary with the substances concerned and with the type of apparatus employed. Joule realized quite clearly that if all the work is transformed into heat then the same quantity of work should produce the same amount of heat whatever substances or apparatus were used. He succeeded in obtaining fairly concordant results in different series of experiments such as (1) Friction of water contained in a brass vessel with a brass paddle, (2) Friction of mercury contained in an iron vessel with an iron paddle, (3) Friction of two iron rings rubbing against each other in mercury.

The fact that the same quantity of work always gives rise to the same quantity of heat is a result that can be included in the great synthetic principle called the conservation of energy. In this principle a new and highly abstract concept, the general concept of " energy," emerges. According to this principle the total amount of energy in the universe is constant. Energy may take new forms, but it cannot be created or destroyed. Thus, as Joule's experiments showed, the energy of a falling weight can be converted into heat. Conversely, heat can be converted into mechanical work, as in the case of a steam-engine. Again, the energy of motion of moving water or moving air can be diverted and made to reappear in another form, as in the case of a water-mill or windmill. The energy of strain stored up in a bended bow appears, when the string is released, as the energy of motion of the flying arrow. Energy that vanishes in one form reappears in another, and the principle asserts that the quantity of energy in the universe, like the quantity of matter, is constant. But in order that the total amount of energy in the universe should remain constant it is necessary to invent the concept of " potential " energy. A stone let fall from a height, for example, acquires velocity and therefore energy of motion. On striking the earth this energy is transformed into heat. But where does this energy come from? It would appear that energy can be created merely by letting a stone fall.

The principle of conservation is preserved, in this case, by saying that the stone, before it fell, possessed " potential " energy. Owing to the relative positions of the stone and the earth the stone has a latent capacity for doing work. Thus a stone on top of a mountain has more energy than an exactly similar stone at the foot of the mountain. As the stone falls it loses potential energy and acquires energy of motion, or kinetic energy. It is the sum of the two energies that is conserved. The amount of kinetic energy gained is equal to the amount of potential energy lost. It must be admitted that this notion of potential energy is likely to seem to the reader somewhat mysterious. The transformation of energy of motion into heat energy seems to be a genuine case of transformation. Both forms of energy are obviously " energetic." But this purely passive potential energy seems to be undetectable except when it becomes something else. We have here an instance of a scientific principle which is partly conventional. Potential energy is a notion brought in to save the principle of the conservation of energy. The potential energy of an isolated system is so defined as to allow the total energy of that system to be conserved.

The merit of the device lies in its usefulness. It is a convenient way of summarizing phenomena, but it should not be given, as it often is, a metaphysical status to which it has no claim. Scientific men are usually somewhat reluctant to admit the conventional character of some scientific principles, but the conventional character of potential energy is clear. This had led some writers, anxious to assert the objective existence of potential energy, to say that all potential energy is really kinetic. For this purpose that universal provider, the ether, is, of course, brought into play. The point of view is well put in a famous treatise on heat: [1]

" The question still remains, what becomes of the motion when the kinetic energy of a system diminishes? Can motion ever be changed into anything else than motion? If we assume a fundamental medium whereby to explain all the phenomena of Nature, then the properties of this medium ought to remain unchanged, and all other changes must be explained by motion of the medium. Such an assumption is quite philosophic, and the method of procedure is certainly scientific. An evident reply to the question of what becomes of the motion of a projectile rising upwards is that it passes into the ether. The first assumed property of the ether is that it can contain and convey energy. There is no *a priori* reason, then, why the energy of motion of a projectile as it rises upwards should not be stored up as energy of motion of the ether between the body and the earth, or elsewhere. The oscillation from kinetic to potential, and from potential to kinetic, in the case of the pendulum is then, from this point of view, merely an interchange of energy of motion going on between the mass of the pendulum and the ether around it. According to this view all energy is energy of motion, and must be measured by the ordinary mechanical stan-

[1] Preston's *Theory of Heat.*

70

dard. The work we do in lifting a body from the earth is spent in generating motion in the ether, and as the body falls this motion passes from the ether to the body, which thus acquires velocity.

" A rough mental picture of the process might be obtained as follows: We might suppose a body connected to the earth by vortex filaments in the ether, which would replace the lines of force. The ether is spinning round these lines, and when the body is lifted from the earth the work done is expended in increasing the length of the vortex filaments. The work is thus being stored up as energy of motion of the ether, and when the body falls to earth the vortex lines diminish in length, and their energy of motion passes into the body and is represented by the kinetic energy of the mass."

It is obvious that these speculations are designed merely to present potential energy as something more than a convenient mathematical fiction. If we reject them we must admit that potential energy, and, to some extent, the principle of the conservation of energy, are merely conventions. In modern relativity theory the notion of potential energy is not admitted. Energy of motion, including molecular motions, is admitted, and this is found to be, not exactly conserved, but something rather like it. The point is interesting as it calls attention to the conventional elements that may enter into scientific generalizations. The physicists of the nineteenth century practically used the principle of the conservation of energy to define energy. Energy became that something that was conserved. In this way the principle was placed beyond the reach of experimental verification. Thus the writer already quoted says that if a system is found to be losing energy in a way we cannot account for we must suppose that it is radiating energy in some unknown manner into the ether. Similarly, if the system is found to be gaining energy we must assume that the ether, that vast store of energy, is somehow conveying energy to it. The conservation of energy thus becomes a completely irrefutable principle.

We see that there is a considerable variety of status possessed by the scientific ideas introduced up to this time. Mass was regarded as a fundamental and irreducible characteristic of matter. The luminiferous ether was also regarded as an actual entity whose properties were in process of being worked out. The electric fluids were merely first approximations to the physical reality that underlay the manifestations of electricity. They were not mere conventions, since nobody doubted that " electricity " existed. But that electricity was actually a fluid was probably not seriously asserted. Energy, as we have seen, was partly a name for a definite characteristic of moving matter, and partly a convention. The principle of the conservation of energy was also partly conventional, since it was immune, by definition, from experimental test. It therefore had an entirely different status from the principle of the conservation of mass. This latter principle was the result of very careful weighing experiments by which it was shown that the quantity of matter concerned in the experiment remained constant, whatever physical or chemical changes it passed through. If the experiments had consistently shown that matter was gained or lost, the principle that matter could not be either created or destroyed would have been given up. The notion of an undetectable matter, a " latent " matter, would not have been brought in to preserve the principle. The reason is that such a generalized notion of matter would have been of very little use, whereas the generalized notion of energy enabled a remarkably convenient description to be given of very diverse phenomena.

The fact that heat can be generated by the expenditure

of work was, as we have seen, demonstrated by Joule. The inverse of this, that work may be performed by the expenditure of an equivalent quantity of heat, was first investigated experimentally by Hirn in 1857. By experimenting on a steam-engine, measuring the amount of work performed and the amount of heat used after allowing for the quantity lost by radiation and conduction over all parts of the machine, Hirn arrived at a very fair estimate of the quantitative relations between heat and work.

The statement that there is a quantitative relation between heat and work is called the First Law of Thermodynamics. A truly scientific formulation of the bases of Thermodynamics was first given by Sadi Carnot in 1824, but his work received no proper attention for something like twenty years. It leads directly to the famous Second Law of Thermodynamics, which was stated by Clausius and Lord Kelvin. The form of the law, as enunciated by Clausius is:

"It is impossible for a self-acting machine, unaided by any external agency, to convey heat from one body to another at a higher temperature, or heat cannot of itself (that is, without compensation) pass from a colder to a warmer body."

Kelvin's form runs:

"It is impossible by means of inanimate material agency to derive mechanical effect from any portion of matter by cooling it below the temperature of the coldest of surrounding objects."

We here meet with a notion which, in its assumptions and ramifications, is of very great importance in science. That is the notion of *reversible and irreversible* phenomena. Elementary natural phenomena are supposed to be reversible, that is, they can work backwards as readily as they can work forwards. The swing of a frictionless pendulum in a vacuum is an ex-

ample of a reversible phenomenon. Given the characteristics of the pendulum at any moment, its past history
can be deduced as well as its future history. This is
characteristic of all " purely mechanical " phenomena.
In giving a mathematical description of these phenomena
we make reference, of course, to Time. Time is one of
the variables that enter into our mathematical equations.
But time, as it occurs in science, does not correspond
to our subjective sense of time. For the time of science
there is no essential distinction between past and future.
The relation to the present moment of past and future is
no more, for science, than the relation of right and left
to a point on a line. In other words, time does not enter
into science as a creative factor. In this the scientific
conception of time is utterly different from the mediaeval
conception, which regarded time as the actualization of
potentiality. If we say, in science, that the past causes
the future, it is only in the very limited sense that the
future can be deduced from the past. Since we can also
deduce the past if we are given the future, we could
equally say that the future causes the past. There are
some doubts, as we shall see later, whether this conception of time is really adequate for modern science. There
are indications that we may have to adopt a conception
more akin to the mediaeval one, and make explicit reference to a *process,* tending towards an *end,* in our description of natural phenomena. The general assumption,
however, is that reversible phenomena are fundamental in
Nature.

Yet many irreversible phenomena occur. Any actual
pendulum, for instance, will sooner or later come to rest.
Through the friction of the supports and the resistance
of the air its motion is gradually dissipated. And obviously,
from examination of a pendulum at rest, its previous
active career cannot be deduced. Similarly a number of

bodies at different temperatures enclosed in a room will all finally reach the same temperature. According to the second law of Thermodynamics this state of affairs will never be spontaneously reversed. And, from the final temperature, the initial distribution of temperature amongst the bodies could not be deduced. All such cases are explained as being merely statistical effects of a large number of reversible phenomena. The actual elementary phenomena involved, such as molecular motions, are all reversible.

A very clear illustration of the considerations involved may be reached if we consider the case of two vessels, one containing say, nitrogen, and the other oxygen, at equal pressures. Let the two vessels be put into communication by a tube that can be closed by means of a turn-cock. If we open the connecting tube the two gases will mingle and remain mingled. We naturally suppose that the initial condition would never return. The final state of mingling is irreversible. Yet all the molecular motions involved are strictly reversible. The actual motion of the system must be periodic and, after a sufficient lapse of time, the initial condition should return. This is, in fact, the case. If our laws of molecular motion are true, then, after a tremendous lapse of time, we should again have nothing but nitrogen in one vessel and nothing but oxygen in the other. In the meantime we have intermediate states of the system and, as there are an enormous number of these all corresponding to the " well-mingled " state, the chances that the system is at any moment in a well-mingled state are enormously high. In this typical case, therefore, an apparently irreversible phenomenon is the result of a large number of reversible phenomena. The general scientific belief is that all irreversible phenomena are of this kind. Irreversibility is therefore a merely relative term, and has reference to our means of operating on the system under consideration.

An example of this is given by Maxwell's famous "demon." According to the laws of thermodynamics it is impossible to produce any inequality of temperature or pressure in a gas kept at the same volume and protected from the passage of heat, without expending work. Now a gas consists of molecules moving in all directions with all velocities. As the temperature increases the average velocity of the molecules increases. Now, in Maxwell's words: "Let us suppose that such a vessel is divided into two portions, A and B, by a division in which there is a small hole, and that a being, who can see the individual molecules, opens and closes this hole, so as to allow only the swifter molecules to pass from A to B, and only the slower ones to pass from B to A. He will thus, without expenditure of work, raise the temperature of B and lower that of A, in contradiction to the second law of thermodynamics." By a similar operation on the part of a similar demon we see that the original distribution of nitrogen and oxygen in the two vessels mentioned above could be restored. And this restoration would be achieved without violating any of the laws of Nature and without the expenditure of work. It would seem, therefore, that phenomena are called irreversible only with respect to practical operations. We cannot, in practice, reverse the motions of the air molecules struck by the swinging pendulum, nor turn into work the heat energy developed by the friction of the supports, but, if we could, the pendulum would swing for ever.

Yet, if we accept this practical criterion of irreversibility, many phenomena usually classed as reversible become irreversible. If we make a solution of gold, for example, the process is called irreversible. Except by expending energy we cannot get the gold back. But if a pound of gold dust be scattered over a wide area so that it is hopelessly mingled with dust and refuse it has suffered a merely mechanical mingling, and is theoretically recoverable without the expenditure of energy. Yet the recovery of the gold dust, in practice, might require a much greater expenditure of energy than would its recovery from a chemical transformation. Such examples of irreversibility are of great importance in the lives of organisms, as A. J. Lotka[1] has pointed out, and he suggests that a mathematical analysis will have to be invented to deal with such cases. The question may be summed up as follows: All irreversible phenomena are merely statistical effects of reversible phenomena if it be true that, for the mathematical description of the elementary operations of Nature, the direction of time is indifferent; that is, that no essential distinction need be made, for the purpose of describing these operations mathematically, between past and future. If *strictly* irreversible phenomena exist, then their mathematical description will require a different notion of time. Time will appear as a creative factor in Nature. The further discussion of this question must be deferred till later.

With the abandonment of the notion of caloric, the ether was called upon to provide the explanation of *radiant* heat. Light rays convey heat as well as light, and heat rays behave, on experiment, very much like light rays. We may suppose that heat rays are invisible light, or light rays visible heat. The difference between them is presumably a difference of wave-length, the heat-waves being the longer. The rapid molecular vibrations that constitute the heat of bodies may be supposed to set up these waves in the ether. If we assume that heat-waves are only long light-waves we may conclude that they travel with the same velocity as light. The theory was perfectly

[1] *Elements of Physical Biology.*

77

well stated by Dr. Thomas Young, the only contemporary of Rumford and Davy who had clear ideas on the subject. He says:

"If heat be not a substance it must be a quality, and this quality can only be motion. It was Newton's opinion that heat consists in a minute vibratory motion of the particles of bodies, and that this motion is communicated through an apparent vacuum by the undulations of an elastic medium, which is also concerned in the phenomena of light. If the arguments which have been lately advanced in favour of the undulating nature of light be deemed valid, there will be still stronger reasons for admitting this doctrine respecting heat, and it will only be necessary to suppose the vibrations and undulations principally constituting it to be larger and stronger than those of light, while at the same time the smaller vibrations of light, and even the blackening rays, derived from still more minute vibrations, may perhaps, when sufficiently condensed, concur in producing the effects of heat. These effects, beginning from the blackening rays, which are invisible, are a little more perceptible in the violet, which still possess but a faint illuminating power; the yellow-green afford the most light; the red give less light, but much more heat; while the still larger and less frequent vibrations, which have no effect on the sense of sight, may be supposed to give rise to the least refrangible rays, and to constitute invisible heat."

The "blackening rays" referred to by Young are rays of too short wave-length to affect our sense of sight, but which can be detected by their chemical effects. We now know that the range of "ether-waves" is vastly greater than was suspected by Young, extending, as they do, from the electromagnetic waves used in wireless telegraphy to the recently discovered "cosmic rays," which are shorter than the shortest X-Rays. But we see that, even a century ago, the two great phenomena of light and heat were linked up, in certain aspects at least, in one great generalization. This generalization was still further generalized, and a great new region won for science, when Maxwell created his electromagnetic theory of light.

CHAPTER V

MOLECULES AND ATOMS

WE have said that the theory that matter is composed of atoms and molecules was first given precise form by Dalton, at the beginning of the nineteenth century. The hypothesis, as we have seen, was a very old one. It existed in a vague form amongst the ancient Greeks. But Dalton was the first man to put it in a form that admitted of quantitative verifications. The theory was a great success in chemistry from the very moment it was enunciated, but it is only in recent times that nobody can be found to say a word against it. Goethe was opposed to it, and we may also mention the more scientific names of Mach and Ostwald. Faraday doubted it. But the theory has now been confirmed with such precision, and in so many different ways, that doubt is no longer possible. It is true that the actual shape and constitution of the atom are matters that still present great difficulties, but as against the theory that matter is continuous the atomic theory has triumphed. A great deal can be done with the atomic theory even on the crude supposition that atoms are small hard spheres. As an illustration of this we may mention, besides its triumphs in elementary chemistry, two of its physical applications, the kinetic theory of gases and the Brownian movement.

If matter is composed of small separate particles we may suppose that, in a solid, the movements of these particles are severely restricted. The motions that constitute heat in a solid we may suppose to consist in small oscillations about a centre. Yet even here we have evidence of actual translatory motions of the molecules. With pairs of metals that have been kept in contact for years, we find that the bottom layer of the top metal and the top layer of the lower metal have become to some extent intermingled. With liquids and gases, of course, diffusion is much more rapid and complete. Thus, if a globe containing hydrogen, the lightest of gases, be placed above a globe containing carbon dioxide, a heavy gas, and the two globes be put in communication by a stopcock, it is found, after a short time, that each globe contains as much carbon dioxide as hydrogen. Liquid diffusion takes longer, but even if ether and water be superposed it is found, after a time, that every part of the layer of ether contains some water and that in every part of the water there is some ether. The maximum freedom of movement, however, is possessed by the molecules of a gas.

On the molecular theory, as we see, we may suppose a gas to consist of an enormous number of particles moving in all directions at random. We may further suppose that some particles are moving more swiftly than others, and that the directions of the particles are continually changing, due to collisions. The kinetic theory of gases is the attempt, inaugurated by Maxwell, to deal mathematically with this chaotic assemblage. It would obviously be hopeless to attempt to follow the history of one particular molecule. Maxwell therefore introduced statistical methods and applied the theory of probabilities. The mere fact that the molecular chaos could be assumed to be completely chaotic was found to provide the key to the problem. In this investigation the molecules were supposed to obey the ordinary laws of mechanics, and on this basis, the well-known laws governing the behaviour of gases were successfully derived from the theory, and certain new phenomena predicted which were experimentally confirmed. There is no need, for our purposes, to pass these results in review, but we may refer to the very useful notion of " absolute temperature " which is made very clear by this theory.

The temperature of a gas is a measure of the kinetic energy of its molecules, and the way in which this energy changes with the temperature was calculated. It changes in such a way that, at a temperature 273 degrees below zero on the centigrade scale, it becomes zero. This, therefore, is the absolute zero of temperature. No body can be colder, since energy cannot assume negative values. The actual velocities of the molecules of a gas, at ordinary temperatures, as derived from this theory, are very considerable. Molecules of air are moving with about the velocity of a rifle bullet. On an average, the distance traversed by each molecule between successive impacts with other molecules is one ten-thousandth of a millimetre. Each molecule experiences about five thousand million collisions per second. The size and mass of these tiny bodies can also be calculated. Thirty thousand million of them placed side by side, would give a length of one centimetre. Twenty thousand million taken together, would have a mass of one thousand-millionth of a milligramme. These estimates are only rough and, of course, molecules of different substances have different sizes and weights, but they illustrate the order of magnitude involved. Molecular velocities also vary, assuming the same temperature, with the kind of molecule involved. Thus a molecule of hydrogen is moving four times as fast as a molecule of oxygen, at the same temperature. The kinetic theory of gases, when it comes to matters of detail, is not without its difficulties, but its success is certainly sufficiently great for it to be regarded as a very strong support of the molecular theory of matter.

Another phenomenon, which may be said to make molecular motions visible, ᵥis the so-called Brownian Movement. An English botanist named Brown, using the improved microscope that had just been introduced, noticed, in 1827, that very small particles suspended in water were in a state of constant agitation. This discovery attracted little attention. It was dismissed as being due to vibration, or the result of slight inequalities of temperature and pressure, as when dust particles dance in a sunbeam. When the motion came to be more particularly observed, however, it was found that these explanations would not work. It was certainly not due to vibration. Gouy found that it occurs just as vigorously, on a fixed support and at night in the country, as in town on a table continually shaken by the passage of heavy traffic. Again, when the greatest care was taken to maintain uniformity of temperature throughout the drop of water, the irregular motion was not diminished. Also it was shown that the phenomenon could not be referred to the influence of light. Changes in the intensity or colour of the light illuminating the drop were without effect. The actual nature of the small particles suspended in the water is without effect. The motion is more violent the smaller the particles, but, provided the particles are of the same size, their substance or density does not affect the motion. It became evident that the Brownian movement is not as trivial a phenomenon as it might seem. The movement never ceases. If proper precautions are taken against evaporation it may be watched for years. It has been observed in liquid inclusions that have remained shut up in quartz for thousands of years.

This movement has now received a thoroughly satisfactory explanation. The theory was worked out by Einstein in 1905, and his quantitative conclusions have been experimentally confirmed by Perrin. We now know that the Brownian movement is due to the molecular agitation of the liquid in which the particles are suspended. The random motions of the fluid molecules, striking the particle on all sides, usually cancel out. But the smaller the particle the less the chance that the irregular impacts, arriving at a given instant, shall cancel out. For a time sufficient to produce an observable motion it will happen, now and then, with a sufficiently small particle, that the impacts have a resultant in one direction. The next moment, of course, the direction has changed. And so we get this incessant and extremely irregular motion of the suspended particle. We have here a magnificent illustration of the molecular theory of matter, and of the fact that the molecules are in a state of constant agitation. The agreement between this theory and actual observation is complete.

It is worth noting that the Brownian movement, like Maxwell's demon, illustrates the fact that the second law of thermodynamics is really a statistical law. It concerns large-scale phenomena and has reference to practical human limitations. We must suppose that the Brownian movement conforms to the law of the conservation of energy. It is not an example of " perpetual motion." But the law of the conservation of energy can only be obeyed in this case if we suppose that every increment of velocity acquired by a suspended particle is accompanied by the cooling of the fluid in its immediate neighbourhood. This is an apparent violation of the second law of thermodynamics. That law states, for instance, that we could not propel a ship by cooling the sea-water in its neighbourhood. The assumption here is that we cannot, practically, take advantage of the actual inequalities of the molecular energies of sea-water, and the assumption is perfectly justified. But on the small scale of the Brownian movement, the suspended particle takes advantage of the fact that a drop of water is not at the same temperature throughout, since different molecules have different kinetic energies, and the kinetic energy of the same molecule changes, from collisions, every instant. Hence on this scale the second law of thermodynamics, which is a statistical law, does not apply. Hence the second law of thermodynamics is only true because we cannot deal practically with magnitudes below a certain limit. If our universe were populated by intelligent bacteria they would have no need of such a law.

As we have seen, the atomic theory throws a great deal of light on phenomena even if we assume no more about the atom than that it is a small hard particle. For a good many chemical theories, also, this conception is sufficient, a little complicated by granting to atoms various " affinities " for one another. But when the atoms were arranged in order of their relative weights the suspicion arose that this conception of the atom was not sufficient. The relative weights of atoms, where an atom is defined as the smallest part of an element that enters into any known chemical process, can be determined with considerable accuracy. When a list of the relative weights of the atoms is formed it is found to extend from hydrogen, the lightest, to uranium, the heaviest. Oxygen, as the most frequently occurring of the elements, is taken as the standard with which the weights of the other elements are compared. An oxygen atom is nearly sixteen times as heavy as a hydrogen atom. For convenience, therefore, its weight is represented by the figure 16 exactly. On this scale the relative weight of the hydrogen atom is 1·008, and the relative weight of the uranium atom is 238·5. Now although no two chemical elements are exactly alike it is nevertheless true that they can be arranged in groups such that the members of each group have similar properties. Thus the elements Lithium, Sodium and Potassium are members of such a group. The elements Beryllium, Magnesium and Calcium belong to another such group. A large number of such groups can be formed.

Now a curious fact about these groups emerges when we number the elements in the order of their atomic weights. We start with hydrogen as number 1 and finish with uranium as number 92. On this scale lithium, sodium and potassium have numbers 3, 11 and 19 respectively. The elements beryllium, magnesium and calcium have numbers 4, 12 and 20. Another group of similar elements has the numbers 5, 13 and 21. Yet another group has the numbers 6, 14 and 22. And so on. We see that similar elements occur at equal intervals as we go down the table. Thus $11-3=19-11=8$. And the interval is the same for the next group. Thus $12-4=20-12=8$. And the same for the other groups. This arrangement was suspected by Newlands in 1864, but it was first put forward convincingly by the Russian chemist Mendeléev about 1870. This is the so-called " periodic system " of arranging the elements. It appears that approximately the same set of chemical properties belongs to each eighth member of the table, starting with any one we like. This rule, however, is too simple. It works well enough for the early part of the table, but later on in the table we find that the recurrence of chemical properties begins after the eighteenth instead of the eighth member, and, later on still, after the thirty-second member. Nevertheless, the recurrence is remarkably regular, and must be very significant. It is difficult to see what it can be significant of, except a resemblance in structure of the similar elements.

On the supposition that atoms are simply small hard spheres of different sizes and weights such a recurrence of properties seems quite inexplicable. If we suppose atoms to possess structure, we may suppose that their chemical properties depend on this structure, and that similar elements have a similar structure, the heavier atoms being more complicated versions of the same ground plan. The very strong suspicion thus aroused, that atoms possess structure, is the most important theoretical consequence of Mendeléev's discovery of the periodic system. Amongst its practical consequences is the fact that the properties of unknown elements can be described before they have been discovered. The numerical position of the gap in that table shows at once to which group the unknown element belongs. Such predictions have often been made and subsequently verified. It must have seemed that unknown elements were robbed of much of their interest. Unprecedented properties were not to be expected. The discovery of radium, however, assured us that Nature still abounds in surprises. Radium, when discovered, exhibited the properties that could be predicted, but also others that no chemist had ever imagined.

Another indication that atoms possess a complex structure is derived from spectrum analysis. The light from any glowing substance, when passed through a prism and spread out on a screen, is seen to be crossed by lines that are entirely typical of the substance. So entirely definite and characteristic are these lines that astronomers have no hesitation in deducing, from the spectrum of a star, the chemical elements that are radiating the light. Now these lines, it can be shown, must be taken as indicative of atomic processes—oscillations or what not. But the varieties of movement that can be attributed to a small hard sphere are very limited. Yet the spectrum of iron, for example, contains thousands of lines. Here again, then, is an indication that atoms possess structure.

CHAPTER VI

ELECTROMAGNETISM

THE fundamental science in modern physics is the science of electricity. Radiant heat, light, and matter itself are among its manifestations. Of the main orthodox divisions of physics, Sound is the only one that can still be pursued in comparative independence of electrical concepts. But sound is, as it were, a local phenomenon, and in the modern investigation of the material universe plays a comparatively insignificant rôle.

Until the last decade of the eighteenth century the only electrical phenomena investigated were those of statical electricity. Towards the end of the century Luigi Galvani, Professor of Anatomy at Bologna, happened to notice that a dissected frog was violently convulsed merely on being touched with a scalpel. This observation led him to undertake a series of experiments to determine exactly under what conditions these convulsions were produced. He found that the limbs of the frog were convulsed whenever a connection between the muscles and nerves was made by means of a metallic arc, the metallic arc usually being composed of different metals. In his own words:

" For, while I with one hand held the prepared frog by the hook fixed in its spinal marrow, so that it stood with its feet on a silver box, and with the other hand touched the lid of the box, or its sides, with any metallic body, I was surprised to see the frog become strongly convulsed every time that I applied this artifice."

Galvani believed these effects to be due to a peculiar

fluid contained in the nervous system and which passed to the muscles by way of the metallic arc. The question then arose, What is the nature of this fluid? One school supposed that *Galvanism,* or *Animal Electricity,* as it was called, was a fluid quite distinct from ordinary electricity. Another school, to which Galvani himself belonged, maintained that the fluid in the nervous system was the ordinary electric fluid. But a third party arose which declared that the effect had nothing to do with a fluid in the nervous system at all, that, in fact, the physiological peculiarities of the frog were quite irrelevant to the phenomena. The frog merely played the rôle of a moist body. Alessandro Volta, the leader of this school, stated that there were two kinds of electric conductors, dry and moist. The placing in contact of these two species of conductors agitated the electric fluid. *How* the agitation was produced he did not profess to know. Another member of this school, Fabroni, decided, on the basis of certain experiments, that the galvanic effect was inseparably connected with chemical action.

The subject was advanced considerably when Volta found a way of intensifying the effects by properly arranging his dry and moist conductors in a chain. Such an arrangement was called a pile, and he found that when connection between the extreme ends of the pile was maintained by the human body, sensations were experienced during the whole time that contact was maintained. He thus arrived at the conception of a continuous electric current. Since he believed that this continuous activity was produced merely by the contact of the different sorts of conductors it appeared, of course, highly mysterious. As he says: "This endless circulation or perpetual motion of the electric fluid may seem paradoxical, and may prove inexplicable; but it is none the less real, and we can, so to speak, touch and handle it." It is worth nothing that Volta and his contemporaries found this phenomena "paradoxical," although these observations were made long before the principle of the conservation of energy was formulated. The "moist conductors" in these experiments were supposed to have no rôle beyond that of being conductors, although the significant fact had been noted that when the moisture was acidified the pile was more efficient. This fact was attributed solely to the superior conducting power of acids.

Knowledge of the electric current was greatly increased at the beginning of the nineteenth century by the discovery that it could produce chemical actions. Thus Nicholson and Carlisle found that it could decompose water into its constituent gases, hydrogen and oxygen. It was also shown that solutions of metallic salts were decomposed by the electric current. It was already known that electricity created by friction, "frictional electricity," could produce chemical effects, and the identity of the effects led to the conclusion that frictional electricity and voltaic electricity were the same thing. The next great step was to prove that voltaic electricity was, as Fabroni had surmised, itself the product of chemical actions. This was the result largely of experiments made by Humphry Davy, and in the course of them he was led to the great generalization that chemical affinity itself is essentially of an electrical nature. He reduced all chemical attraction to electrical attraction. He says: "Chemical and electrical attractions are produced by the same cause, acting in one case on particles, in the other on masses of matter; and the same property, under different modifications, is the cause of all the phenomena exhibited by different voltaic combinations." This generalization is another example of the "inspired guessing" to which great men of science are prone. It was not justified by the evidence, and even now, although no one doubts it, it cannot be said to be proved.

The discovery of the chemical origin of the electric current had for its chief theoretical consequence the removal of these phenomena from the realm of pure magic, so that they were later able to be included in the generalization of the conservation of energy. But these investigations, although they made electric phenomena an integral part of the general body of scientific knowledge, and although they led Davy, as we have seen, to far-reaching speculations on the essentially electrical character of chemical phenomena, threw light on only a limited region of the activity of this protean and universal agent. The next great step was made by establishing a connection between electric and magnetic phenomena.

That such a connection exists had long been surmised. That lightning could magnetize steel, for instance, had been known since the early half of the eighteenth century. A paper published in the *Philosophical Transactions* of 1735, reports that a tradesman named Wakefield, " having put up a great number of knives and forks in a large box, and having placed the box in the corner of a large room, there happened in July, 1731, a sudden storm of thunder, lightning, etc., by which the corner of the room was damaged, the box split, and a good many knives and forks melted, the sheaths being untouched. The owner emptying the box upon a Counter where some Nails lay, the Persons who took up the knives that lay upon the Nails, observed that the knives took up the Nails." That thunderstorms have an effect on the magnetic needle was also known, and unsuccessful experiments were made to determine whether a voltaic pile, freely suspended, was influenced by terrestrial magnetism.

It was not until 1820 that the first public description was given of the influence of an electric current on a magnetic needle. Oersted was the discoverer of the fact that if a wire carrying an electric current be placed parallel to a magnetic needle the needle is markedly deflected. It was thus proved that an electric current creates a magnetic field, and the question arose whether an electric current would behave like a magnet in a magnetic field. That it does do so was proved experimentally by Oersted. Further properties of the electric current, regarded as a magnet, were obtained by Ampère, who showed, a week after the news of Oersted's first discovery reached the French Academy, that two parallel wires carrying currents attract each other if the currents are flowing in the same direction, but repel one another if the currents are flowing in opposite directions. Oersted himself did not give the quantitative laws of any of these phenomena. He explained them, vaguely, by a theory reminiscent of the magnetic views of Descartes. He supposed an activity, that he calls the " conflict of electricity " in the surrounding space. "All non-magnetic bodies appear penetrable by the electric conflict, while magnetic bodies, or rather their magnetic particles, resist the passage of this conflict. Hence they can be moved by the impetus of the contending powers." This explanation has the merit that it refers the effects to the surrounding space. It is an attempt at a physical theory, as distinguished from the purely formal mathematical theory that was to be elaborated by Ampère. We have already referred to Ampère's celebrated memoir, published in 1825. Physical hypotheses, as we have said, play very little part in it. The magnetic force exerted by an electric circuit is given precise expression, and Ampère shows what sort of magnetized body would be equivalent in its magnetic effects to a given electric circuit. But Ampère did not regard the magnetic fluid as a fundamental entity; it was, according to him, an electric phenomenon. He put forward the hypothesis, which has persisted, that a magnetic molecule owes its magnetism to the fact that an electric current is perpetually flowing in a small closed circuit within the molecule.

The French mathematicians, as we have seen, had evolved a mathematical analysis peculiarly fitted to deal with " action at a distance " phenomena. A characteristic of their method was that it ignored whatever processes may have been supposed to go on in the space separating their " attracting " or " repelling " bodies. Another assumption, implicit in the equations, was that any such influence was transmitted instantaneously. The simplicity and clarity of these assumptions made them extremely congenial to the mathematical mind and, for a large part of the nineteenth century, the mathematical investigation of electrical phenomena was pursued in these terms. Contemporary with this outlook, however, was an entirely different one elaborated by that greatest of all experimentalists, Michael Faraday. Faraday may fairly be called a mathematical genius, although he had no mathematical training and could not use mathematical symbols. He had an extraordinary capacity, however, for isolating, in a complex phenomenon, those aspects between which quantitative relations existed. That quite indefinable " sense of physical reality " that Newton possessed and Leibniz did not, that Maxwell possessed and Cauchy did not, was very highly developed in Faraday. He nearly always attacked a problem by the right end and on the basis of the right assumptions, even when those assumptions were of a quite revolutionary description.

If a sheet of paper be placed over a magnet and iron filings be sprinkled over the paper, the filings will be found to arrange themselves in pleasing and regular curves, each curve being linked to the magnet in such a way that it can be regarded as entering it at one point and leaving it at another. Scientific men had long been familiar with this phenomenon. It is described in a scientific journal of 1717, and Petrus Peregrinus investigated a similar phenomenon in the thirteenth century. This phenomenon suggested to Faraday his conception of *lines of force*. The force exerted by a magnet has, at every point of the surrounding space, a certain direction. We may, therefore, imagine curves drawn in the surrounding space and such that, at any point, the direction of the curve at that point is the direction of the magnetic force at that point. These curves are Faraday's lines of force. The curves in which the iron filings arrange themselves correspond to such lines of force, although the representation is imperfect since the iron filings cannot leave the plane of the paper. The conception was further elaborated by Faraday to take into account not only the direction but also the intensity of the magnetic force at any point. For this purpose he replaced his lines by *tubes* of force. Each tube was constituted, as it were, by a sheaf of lines of force. At places where the force was weak the tube was wide; at places where the force was intense the tube was narrow. The decrease in intensity was compensated for by the increase in the area of the tube. Thus for every tube of force the actual amount of force comprised within its boundaries was constant all along the tube. The intensity of the force at any point, multiplied by the area of cross section of the tube at that point, gave a quantity that remained constant all along the tube. In imagining his tubes of force Faraday made them all equal in the sense that they all contained the same amount of force. Tubes that penetrated into regions distant from the magnet, that is, to regions where the force was weak, would therefore be wider than tubes all of whose paths lay near the magnet. For simplicity each of these tubes could be called a unit line of force, and the strength of the magnetic field in different places is then represented by the separation or concentration of these unit lines of force. Each line of force is a closed curve which, at some part of its course, passes through the magnet to which it belongs. They are also, in their totality, supposed, of course, to fill the whole of space.

These conceptions, which are not at all vague but, on the contrary, perfectly precise and quantitative, did not appeal to the mathematicians of the time. Faraday, as we have said, could not give these conceptions their correct mathematical formulation. That task was reserved for Maxwell. In the meantime the mathematicians preferred to go on with the action at a distance conceptions. Thus the Astronomer Royal, Sir George Airy, an acute mathematician, speaking of the agreement between observations and the calculations based on the action at a distance theory, says: "I declare that I can hardly imagine anyone who practically and numerically knows this agreement to hesitate an instant in the choice between this simple and precise action, on the one hand, and anything so vague and varying as lines of force, on the other hand."

Faraday's rejection of the action at a distance outlook was probably not based on purely philosophical objections. His rejection of that outlook was more probably due merely to the fact that he could not use it. His very strong pictorial imagination was given nothing to grasp in the action at a distance theory. Undoubtedly, also, Faraday's sense of physical reality was shocked by what appeared to be the pure magic of action at a distance. He could not help picturing this action as being due to processes going on in a medium. He could not imagine an iron filing approaching a magnet of its own free will, as it were. But this was merely the expression of a personal predilection. It might quite well be that action at a distance is an ultimate fact about the universe. The feeling that action at a distance is inconceivable comes from a liking for mechanism and from the possession of a pictorial imagination. Nevertheless, we can say that Faraday's conceptions were nearer the truth than those of the older theory, since, when mathematically formulated, they successfully describe a much greater range of phenomena. But it is only the mathematical formulation that we may, with any real confidence, regard as " true." It is the skeleton—Maxwell's equations—that we actually test by our observations. We may be dubious about any particular body with which it is proposed to clothe the skeleton. As a matter of fact, no perfectly fitting body of " physical reality " has ever been suggested for Maxwell's equations. We may conclude, therefore, that Faraday's successful employment of his lines of force was due to the correspondence between their formal properties and the formal aspects of the phenomena investigated. It may be remarked, incidentally, that it is now beginning to be realized that this is all that can be claimed for any scientific theory.

Faraday seems to have invented his lines of force in order to explain to himself the phenomena of electromagnetic induction, of which he was the discoverer. This discovery was the outcome, he tells us, of an attempt to establish analogies between the behaviour of electricity in motion as currents through conductors, and the behaviour of electricity at rest on conductors. It was well-known, of course, that a charge of static electricity on a body had the power of " inducing " opposite electrical states on bodies in its neighbourhood. It occurred to Faraday that electric currents might possess a similar power. He wondered whether a current flowing in a circuit would induce a current in a neighbouring circuit. He found that such a current was induced, but only at the instant of starting or stopping the first current. Faraday concluded that it was not the mere existence of the inducing current that produced the induced current, but its variation. At the instant when the inducing current was growing from zero to its full strength it created a momentary current in the neighbouring wire. Similarly, at the instant that the inducing current was passing out of existence, dropping from its full strength to zero, it created another momentary current in the neighbouring wire. To test his idea that it is solely the *variation* of the inducing current that produces the induced current Faraday made a further series of experiments. He found that by altering the strength of the inducing current he could produce an induced current. Also, if he moved a circuit about in the neighbourhood of a circuit carrying a current, a current was induced in the moving circuit. Magnets moving in the neighbourhood of a circuit induced currents in the circuit.

Faraday's conception of lines of force served admirably to explain all these phenomena. In each case the induced current appears when the circuit in which it flows is cutting, or being cut by, lines of magnetic force. An electric current is surrounded, as Oersted proved, by circular lines of magnetic force. These circles are at right angles to the wire carrying the current. As the current grows towards its full strength these lines increase in number, cutting across any adjacent conductor. Similarly, as the current dies away and the lines collapse, they again cut across any conductor in the neighbourhood. In all the other cases the same fundamental condition is fulfilled; there is relative motion between lines of magnetic force, whether they are due to an electric current or to a magnet, and the circuit in which the induced current appears. The quantitative laws of the phenomena are given by Faraday as follows: "Whether the wire moves directly or obliquely across the lines of force, in one direction or another, it sums up the amount of the forces represented by the lines it has crossed." "The quantity of electricity thrown into a current is directly as the number of curves intersected."

Suppose, now, that instead of having as our circuit a single turn of wire, we have a single continuous circuit formed of several turns close together. Then it is evident that the line of force due to any one turn threads all the others. If the circuit be suddenly broken, therefore, so that the current stops, each coil is cut by lines of force due to the others and an induced current should be created. Such actually is the case, and to this phenomenon the name of "self-induction" has been given. These phenomena of electromagnetic induction may be described as the cardinal discoveries of electric science. The great bulk of the manufactured electrical energy in the world depends upon them, and their mathematical formulation furnishes the equations that lie at the very heart of electrical theory.

The mathematicians of the old school, as we have said, did not work with Faraday's lines of force, and they explained the phenomena of electromagnetic induction by using the analytical methods with which they were familiar. In this connection we may mention Weber's theory, which is called the first of the *electron* theories. It is so called because it attributes electrodynamic phenomena to the action of moving electric charges. Further, the force between these charges (whether of attraction or of repulsion) is supposed to depend not only on their positions at the moment under consideration, but also on their velocities at that moment.

In electrostatics, as in gravitation, the velocities of the bodies concerned are supposed to have no influence on the forces exerted by the bodies on one another. In

Weber's theory of electric currents the velocities play an essential part. Attempts were made to extend this new idea of Weber's to the phenomena of gravitation. Newton's simple law takes account only of the relative positions of the gravitating bodies. Further, it assumes action at a distance. If this simple law be changed, and a propagation of gravitation with the velocity of light be assumed, the calculated motions of the attracting bodies would be, in some cases, sensibly different from those deduced from the Newtonian law. A striking instance is to be found in the motion of the planet Mercury. The point of perihelion of this planet, that is, the point of its elliptic orbit that is nearest to the sun, should, according to the Newtonian law, gradually rotate in the direction of the planet's motion under the influence of the other planets. The *observed* motion of perihelion is appreciably greater than that calculable on the Newtonian law. Various attempts were made to explain this discrepancy, but none of them were satisfactory. It was hoped, therefore, that the new law of gravitation, modified on Weber's lines, would issue in results in complete agreement with those of observation. The calculation was made, and it was found that only three-eighths of the discrepancy could be accounted for in this way. Another law, of very much the same kind, accounted for three-fourths of the discrepancy. These results were not sufficiently good to encourage astronomers to persevere with these modifications of Newton's law. One of the more remarkable results of Einstein's general theory of relativity is that it gives a complete explanation of the whole matter.

Faraday's conception of lines of force applied not only to magnetic force but also to electric force. On a sphere containing a charge of electricity, for instance, the lines of electric force emerge radially in all directions.

The presence of another body causes the lines which would normally pass near it to bend on to it. Faraday imagined that along his lines of force there is a tension tending to shorten them and also that there is a repulsion between neighbouring lines of force. All the phenomena of electrical attraction and repulsion can be given graphical representation by these means. These conceptions, also, were given their mathematical formulation by Maxwell.

Maxwell's electromagnetic theory, the greatest physical theory of the nineteenth century, had its origin in the mechanical concepts elaborated since the time of Newton. The assumption that underlay his work, and all the work of his time, was that the notions of mechanics formed the fundamental categories within which natural phenomena must be arranged. The actual outcome of his work was to destroy the primacy that was attributed to these concepts. Maxwell himself, however, probably never abandoned the belief that electromagnetic phenomena were the outcome of mechanical processes.

Maxwell was the one mathematician of his time who took Faraday's lines of force with complete seriousness. In reading Faraday's *Experimental Researches* he was reading the productions of an imagination very like his own. It was in 1856, when Maxwell was twenty-five, that he gave his first interpretation of Faraday's lines of force. For this purpose he chose the mechanical analogy offered by the lines of flow of a liquid. We have seen that, in Faraday's tubes of magnetic force, the magnitude of the force is everywhere inversely proportional to the cross-section of the tube. If, in a moving incompressible fluid, we construct " tubes of velocity," these tubes will have the same characteristics in this respect as Faraday's lines of magnetic force. For mathematical purposes, therefore, we may represent magnetic force as the velocity of an incompressible fluid. But this, to a mathematician bent on reducing electromagnetic phenomena to mechanism, suggests that the analogy may be physical, and not merely mathematical. Some such analogy, regarded as a physical reality, had previously suggested itself to Faraday when he said, " the physical lines of magnetic force are currents." We may imagine that some ethereal motion is the mechanical cause of magnetic force. The same reasoning enables us to picture the electric force as the velocity of an incompressible fluid, although here the analogy is somewhat less simple. Whatever may be thought of such mechanical analogies, the mathematical treatment is important if only for distinguishing clearly between kindred ideas. Thus the notion of magnetic *induction* has to be distinguished from the notion of magnetic force, and the notion of electric *displacement* from the notion of electric force. In free ether, as distinguished from matter, these distinctions vanish, but they must be preserved when matter is present. It is then found that Faraday's lines represent the magnetic induction and the electric displacement respectively, and not the forces. The reader need not be hampered by these distinctions, although they correspond to physical facts and are of great importance for the mathematical theory.

In 1861–2 Maxwell brought out an investigation where the mechanical explanation of electromagnetism is pushed much farther. In the course of this investigation the fundamental Maxwellian equations appear, but they emerge from a mechanical jungle that is truly extraordinary. Since writing his earlier paper Maxwell had become convinced that magnetism is in its nature a rotatory phenomenon. As he says: " The transference of electrolytes in fixed directions by the electric current, and the rotation of polarized light in fixed directions by magnetic force, are the facts the consideration of which has induced me to regard magnetism as a phenomenon of rotation, and electric currents as phenomena of translation." In order to introduce rotation, therefore, Maxwell imagined that, in any magnetic field, the ether was in rotation about the lines of magnetic force. The centrifugal force so developed would account for the tendency, noticed by Faraday, for the tubes of force to contract longitudinally, and to expand laterally. The right expression for the energy of the field could be obtained if the magnetic force was identified with the circumferential velocity of the etheric vortices. But here a mechanical difficulty arises. Two neighbouring vortices, rotating in the same direction, obviously destroy one another's motion at the place where they touch. In a machine, if it be desired to have two wheels rotating in the same direction, an "idle wheel" is inserted between them. Maxwell therefore supposed a layer of particles, acting as idle wheels, between each vortex and the next. The mathematics of this arrangement was, he showed, in agreement with the known laws of electrodynamics. The motions of the particles represent an electric current. By making this mechanical model still more complicated it could also be made to represent, as Maxwell showed, the phenomena of electrostatics.

So far, it might be thought, we have nothing but an ingenious and complicated model of electrical processes. It does not follow that nothing can be deduced from it. From the identity of the formal relations obeyed by different classes of phenomena we can argue as to the identity of the consequences of those relations. How far Maxwell considered his model to be an actual picture of electromagnetic processes we do not know. Probably he took it quite seriously. At any rate he did not hesitate to draw certain conclusions from his model and apply them to the electromagnetic field. In particular, he was interested to determine the rate of propagation of disturbances through his complex of vortices and particles. He would thus, if his model was sufficiently sound, have solved the first-class problem of the propagation of electromagnetic disturbances. He found that disturbances would be propagated through his model by waves very similar to light waves. Further, he found an expression for the velocity of these waves. The value of that expression had been determined experimentally, and it was found to be the same as the velocity of light. Here, obviously, was a discovery of the very greatest importance. It was legitimate to conclude, and Maxwell did not hesitate to conclude, that light waves are electromagnetic waves, that light is an electromagnetic phenomenon. Here, then, on the basis of what seems to us a very tortuous process of reasoning, was announced the greatest physical discovery of the nineteenth century—indeed, the greatest since Newton.

Maxwell's investigation was not actually as straightforward as we have indicated. In the course of it he had to introduce entirely new ideas, ideas which seem to occur only in those moments of illumination peculiar to geniuses. He asserted, for instance, that all currents are closed currents. According to the ordinary view the current employed in charging a condenser, for instance, is not a closed current. It terminates at the plates of the condenser, where charges are accumulating. Between the plates of the condenser no current is flowing. But Maxwell said that a current was flowing there, and that it produced the same magnetic effects as a true current. His contemporaries, naturally enough, found this a very hard saying. But this apparently mystical current was essential to Maxwell's theory, a fact that long delayed the general acceptance of that theory.

In later writings Maxwell presented his theory with

greater purity, that is, more as the mathematical formulation of the laws connecting electric and magnetic phenomena. Thus, his whole theory may be deduced from two primary equations, one expressing the relation between magnetism and moving electricity, and the other the relation between electricity and moving magnetism. These may be called the laws of Ampère and of Faraday respectively. The actual reasoning by which Maxwell was led to formulate his equations is of no more than historical interest. The justification of the equations does not lie in that reasoning. It lies in their consequences. They have proved completely successful in prediction and have opened up an entirely new world of phenomena.

The experimental verification of Maxwell's theory was not obtained for more than twenty years after its formulation. In the meantime a certain amount of work was done on the basis of Maxwell's ideas. A certain amount of support was given to his theory by John Kerr's experimental demonstration, in 1875, that an electric field could influence the propagation of light. Most of the advances made during this period, however, were made by the mathematicians. Thus J. J. Thomson, in 1881, undertook the examination of the phenomena accompanying a moving electric charge. For this purpose he considered a charged spherical conductor moving in a straight line. This investigation was defective, but it paved the way to a more thorough treatment of the problem. From this work the highly interesting fact emerged that more work is required to communicate a given velocity to a charged sphere than to an uncharged sphere. An electric charge acts, in fact, as if it possessed inertia. This discovery is of great importance, for it suggests that the inertia of ordinary matter may be electrical in origin, and thus clears the ground for an electrical theory of matter. Another theorem of great interest was discovered by Poynting in 1884. According to the older theories the energy of an electric current is carried along in the wire in which the current is flowing. According to Maxwell it exists in the surrounding medium. Poynting worked out the direction and magnitude of the streaming of this energy. In the particular case of a straight wire carrying a continuous current, for instance, the electric and magnetic energy streams into the wire from a direction at right angles to it, and is there transformed into other forms, such as heat energy. It is this transformation of energy streaming from the surrounding medium that constitutes the electric current.

These theoretical researches, important as they were, were completely overshadowed by the experimental verification of Maxwell's theory given by Hertz in a series of investigations starting in 1886. His first experiments were directed towards establishing the fact of propagation through space of electromagnetic disturbances. For this purpose he designed a radiator, a sort of very modified form of Leyden jar, and a detector. His detector was simply a wire bent into an incompletely closed curve, the two ends being furnished with knobs. The radiator gave oscillations of fairly high frequency. When, now, the detector was placed at a distance from the radiator, there being no physical connection between them, and oscillations were excited in the radiator, sparks were observed to pass between the knobs of the detector. It was thus demonstrated that electromagnetic disturbances are propagated through space. Further experiments showed that the velocity of propagation was finite, and of the same order as that of light. A cardinal feature of Maxwell's theory could therefore be taken as proved. Electromagnetic disturbances are propagated through space with the velocity of light. The next step was to show that these disturbances are waves analogous to light waves. Here, again, Hertz was completely successful. He succeeded in reflecting electromagnetic waves from the walls of his room, and also from metal sheets, the law of reflection being the same as that for light. Also, by using a large prism made of hard pitch, he succeeded in refracting electromagnetic waves. In the same year Lodge and Howard found, by passing the electric radiation through large lenses, that it could be concentrated like light. Other experiments established still further parallels between the behaviour of electromagnetic radiations and those of light, so that it may be said that, as the outcome of these experiments, the electromagnetic theory of light was definitely proved.

The mathematical description of Nature made an immense advance with Maxwell's theory, the greatest single advance since Newton's formulation of the laws of dynamics and gravitation. But in doing so it raised a question referring to the very foundation of the scientific outlook. The scientific adventure, as we have seen, had become an attempt to describe the whole of phenomena in terms of certain fundamental concepts that had been abstracted from the observed behaviour of material bodies. The phenomena of light had hitherto been explained, although not with complete success, in terms of these concepts, by postulating the existence of an elastic solid, called the ether, filling all space. Maxwell's equations, as we have seen, accurately described the behaviour of electromagnetic phenomena. But although Maxwell had, as an historical fact, arrived at his equations by thinking in mechanical terms, he had made jumps, and the theory, in its developed form, had no satisfactory mechanical basis. The question was, could such a basis be supplied? In other words, could a description of electromagnetic phenomena be given in terms of the Newtonian conceptions and certain later conceptions that grew out of them, such as the conservation of energy?

It might be, of course, that the mechanical conceptions were inadequate to deal with this new region of phenomena. It had been brought within the realm of mathematics; it had not been brought within the realm of mechanics. But to many physicists of the nineteenth century a mathematical description which established relations between concepts which were not mechanical was thereby incomplete. They could not feel satisfied with such descriptions. Lord Kelvin, a great and representative member of this school, went so far as to say that he could understand nothing of which he could not make a mechanical model. This passion for mechanical explanations was doubtless, to some extent, the result merely of mental inertia. Men, when their minds are mature, are reluctant to think except in terms to which they are accustomed. No *a priori* reason could be given for supposing that the Newtonian conceptions must be adequate for the description of the whole material universe. But the attitude, even if, as a psychological fact, it was largely based on prejudice, could be justified by the axiom that entities are not to be multiplied unnecessarily. Men were certainly justified in seeing whether or no electromagnetism could be reduced to mechanics. But they were hardly justified in rejecting Maxwell's theory, as Lord Kelvin did, because they could not give a mechanical explanation of it.

The attempt to reduce electromagnetism to mechanics took the form of designing ether models. We have already seen that the ether theory had never given a completely satisfactory account of all the phenomena of light. Now that all radiation was regarded as electromagnetic, the ether was required to have such mechanical properties as would issue in Maxwell's equations. An extraordinary number and variety of ethers arose in response to this demand. Some were designed to exhibit electricity as a linear, and magnetism as a rotatory phenomenon. Others inverted the rôles, making magnetism linear and electricity rotatory. Some regarded the ether as an elastic solid; others abandoned the elastic solid analogy. Ethers were then invented which were not continuous media at all, but were complicated conglomerations of vortices. A refinement on these was invented by Kelvin, the so-called " vortex-sponge " ether. All these ethers were successful in representing certain features of the phenomena to be explained, but none of them were completely successful. The one impression they all concur in giving to the reader is that they are incredibly complicated. This fact led some mathematicians, particularly in France, to wonder what their authors intended by these models. Suppose, for instance, that an ether-model consisting of wheels, each wheel geared to four neighbours by india-rubber bands, the bands being of varying elasticity and also capable of slipping, suppose that such a model is shown to be capable of transmitting vibrations analogous to those of light, what precisely is proved? Is it to be supposed that there is a vast medium, filling interstellar space, constituted in this way? Or is it to be taken merely as a *tour de force,* an exhibition of human ingenuity, but corresponding to nothing in natural processes? It seems that the question will be answered differently according to whether one has a mystical conviction that the physical universe is a machine that could be reproduced on a small scale by a nineteenth-century engineer in his workshop, or whether one believes that the concepts of mechanics are merely the first step in the endeavour to isolate abstractions that shall be sufficient for the mathematical description of Nature. In any case, the attempt to give a mechanical description of electromagnetic phenomena has not hitherto been a success. As a consequence of this failure the inverse question has been propounded. Can an electrodynamic explanation be given of mechanics? Or must we regard the electromagnetic concepts as additional to those of mechanics, the complete science of physics using both groups? We shall see later what answer may be given to this question.

CHAPTER VII

THE ATOM OF ELECTRICITY

THE Faraday-Maxwell way of regarding electromagnetic phenomena laid stress upon the processes going on in the *field*. On the basis of that theory Poynting, as we have seen, obtained the energy accompanying an electric current from the surrounding space. Instead of the old action at a distance, the new theory rested upon equations that traced the electromagnetic actions from one instant of time to the next and from one point of space to the next. This way of regarding phenomena made the problem of the ether acute. The centre of interest, as it were, was transferred from matter to space. One consequence of this way of regarding phenomena, as we shall see, was Einstein's theory of relativity. But, side by side with the development of Maxwell's theory, an entirely different aspect of electrical phenomena was being investigated that was to lead to generalizations of equal importance.

We have already seen, from the experiments of Nicholson and Carlisle and of Davy, that when an electric current passes through a liquid conductor the liquid conductor is decomposed. This phenomenon was investigated, by Faraday, amongst others, and, besides determining certain very important quantitative laws, he invented, for this class of phenomena, a terminology that has lasted to this day. Thus the metallic plates through which the current enters and leaves the liquid conductor are called electrodes. The liquid conductor is called an electrolyte, and the process of decomposition is called electrolysis.

Since the very beginning of the nineteenth century attempts were made to frame theories that should satisfactorily account for the phenomena of electrolysis. The earliest of these was advanced by Grothuss and by Davy. They supposed that the electric force applied to the liquid conductor dissociated its molecules into their constituent atoms, these atoms carrying opposite electrical charges. This decomposing force is strongest in the neighbourhood of the electrodes. The positive electrode attracts the negatively charged atoms and repels those having a positive charge. The negative electrode behaves similarly, attracting positive atoms and repelling negative ones. In each case the repelled atoms attack their neighbours and decompose them. The atoms freed in this way attack, in turn, their neighbours, and so on. In this way a chain of decompositions and recompositions travels from each electrode to the other, the atoms that reach the electrodes being given off there. These free atoms are known as ions.

There are various difficulties in the way of this theory. Faraday established the law that the quantity of a substance which is resolved by an electric current is proportional to the current strength. But according to the hypothesis of Grothuss and Davy we should expect that doubling the current strength would quadruple the electrolysis. For the electric force is supposed to dissociate the molecules into ions and also to set the ions moving towards the electrodes. Doubling the force, therefore, should double the dissociation and also double the velocity of the ions. Hence four times the number of ions, in a given time, should appear at the electrodes. This deduction is not confirmed by observation. Again, it would be expected that a certain minimum force would be necessary to effect the decomposition, and that therefore no electrolysis at all would take place unless the force had reached a certain value. This result, also, is not confirmed by experiment.

The subject of electrolysis was studied for several years in considerable detail, but a satisfactory theory was not formulated until 1887, when a young Swedish physicist, Svante Arrhenius, put forward his theory of solutions. He put forward the revolutionary idea that, in very dilute solutions, the electrolyte exists in the form of free ions. Thus common salt, which is sodium chloride, does not exist, in a very dilute solution, as salt. It exists as sodium and chlorine in the form of free atoms. In more concentrated solutions the salt is less completely dissociated. Arrhenius's theory was accepted only reluctantly. His contemporaries found it difficult to believe that there is no salt at all in a solution of salt, but only dissociated atoms of sodium and chlorine. Nevertheless this apparently absurd idea explained so many phenomena that it was finally accepted. In a solution, then, we have free atoms. Each atom of the same kind, for example, all the sodium atoms, carries exactly the same electric charge. This charge must be the same in magnitude, but opposite in kind, to the charge carried by a chlorine atom, since the solution is electrically neutral. When placed in an electric field the positively charged atoms, the positive ions, will move towards the negative electrode and the negative ions will move towards the positive electrode. Thus the electric current in such a solution consists of a double stream of matter in two opposite directions. On coming in contact with the electrodes the ions lose their electric charge.

We have seen that, in a solution of common salt, the charge on the sodium atoms is equal and opposite to the charge on the chlorine atoms. It is reasonable to suppose that, in any other solution containing chlorine atoms, those atoms would have that same charge. If we had a solution of potassium chloride, for instance, we may suppose that the chlorine atoms have the same charge as the chlorine atoms in a solution of sodium chloride. This is made reasonable by the fact that the chemical properties of the two solutions, due to the presence of the chlorine ions, are the same. Since the potassium chloride solution, also, is electrically neutral, the charge on the potassium ions, like that on the sodium ions, must be equal and opposite to that on the chlorine ions. In a molecule of barium chloride, however, we have two atoms of chlorine united with one of barium. In a solution of this substance, therefore, we have two free chlorine ions for every one of barium. The solution being electrically neutral, the charge on a barium ion must be opposite to, but twice as great as, the charge on a chlorine ion. We thus arrive at the conception, which had been reached by Faraday, of a *minimum electric charge*, of an *atom* of electricity. Amongst the charges carried by different atoms, chlorine, barium, lanthanum (whose charge is three times as great as that of a chlorine atom), etc., there must be a minimum charge. The charge of any other atom must either be equal to this (in magnitude) or some whole multiple of it. It may therefore justly be called an atom of electricity. From the phenomena of electrolysis, therefore, we get our first glimpse of the highly important fact that electricity, like matter, may be regarded as having an atomic constitution. The verification of this idea, however, comes from the study of electric conduction through gases.

A gas, in its normal state, hardly conducts electricity at all. It can, however, be made conducting by subjecting it to certain influences. Thus, in the neighbourhood of a red-hot body a gas conducts quite well. The gaseous products from a red-hot flame are also good conductors. And we know now that gas can be made conducting by passing X-rays through it. It is natural to suppose that we are here concerned with the same phenomenon that occurs in electrolysis. We may suppose that the X-rays, or the red-hot body, dissociates some of the molecules of the gas into ions and that, under the influence of an electric force, we get a double drift of these ions, as in electrolysis. This view was quite clearly enunciated as early as 1882, by Giese, of Berlin. He says: "It is assumed that in electrolytes, even before the application of an external electromotive force, there are present atoms or atomic groups—the ions, as they are called—which originate when the molecules dissociate; by these the passage of electricity through the liquid is effected, for they are set in motion by the electric field and carry their charges with them. We shall now extend this hypothesis by assuming that in gases also the property of conductivity is due to the presence of ions. Such ions may be supposed to exist in small numbers in all gases at the ordinary temperature and pressure; and as the temperature rises their numbers will increase." Indeed, the general analogy of conduction in liquids and in gases had long been known. Adolphe Perrot, in 1861, had noticed that when steam is split up into a mixture of hydrogen and oxygen by the passage of an electric discharge through it, there was a decided preponderance of hydrogen grouped about one electrode and of oxygen about the other. But the satisfactory investigation of the whole matter belongs to quite recent times and has definitely established the existence of the elementary electric charge. The crux of the whole matter is, of course, to determine the charge carried by a single ion. Various experiments have been performed to this end, but it will be sufficient to give a description of what is perhaps the most brilliant of them, the experiment due to C. T. R. Wilson and J. J. Thomson.

114

THE ATOM OF ELECTRICITY

It is well known that the amount of water vapour that can exist in a gas depends on the temperature of the gas, being greater the warmer the gas. If a gas, saturated with water vapour, be cooled, a certain amount of the water vapour is deposited in the form of mist or rain. It was found that this deposition of water depends on the presence, in the gas, of small dust particles or solid particles of some sort. The superfluous water vapour condenses in little drops of water round these particles. If the gas be freed from dust by being filtered through cotton-wool it can be cooled to a much lower temperature without condensation taking place. The first step in the experiment we are describing was made when C. T. R. Wilson discovered that ions present in a gas can play the part of dust particles. The ions act as nuclei for condensation. On the basis of this discovery, J. J. Thomson invented an experiment to determine the charge carried by a single ion. By cooling a saturated gas in which ions were present he formed a cloud of water drops round the ions. This cloud began to fall, because the drops were heavier than the gas surrounding them. But such drops, it can be shown, soon attain a constant velocity of descent, owing to the viscosity of the gas. The formula for this velocity had been worked out by Sir George Stokes. In this particular application it involves, besides known quantities, such as the force acting on the drops and the viscosity of the medium, an unknown quantity, namely, the radius of a drop. (The drops are assumed to be spherical and of the same size.) But since, in this experiment, the velocity of descent could be observed, the formula could be used to calculate the radius of a drop. Knowing the radius of a drop and knowing the density of water, the weight of a drop could be calculated. Further, the experiment enabled the total weight of the cloud to be determined. If we neglect the weight of the ions within the drops as too small to affect the result we have only to divide the weight of the cloud by the weight of a single drop in order to obtain the number of drops. The total electric charge carried by the cloud could also be measured experimentally. This charge is due to the ions and the number of the ions is equal to the number of drops. But we have already found the number of drops. If, therefore, we divide the total charge by this number we reach the value of the charge carried by each ion. Thus we arrive at the magnitude of the atom of electricity.

This experiment was repeated, in an improved form, by H. A. Wilson. He reached the same figure as Thomson for the magnitude of the elementary charge, but whereas Thomson's experiment measured the average charge on both negative and positive ions, Wilson's form of it measured the charge on negative ions only. The coincidence of the results proved therefore that the positive and negative ions carry charges opposite in sign but equal in magnitude. Refinements on these experiments have enabled individual drops to be observed. Thus Millikan has formed drops, of a non-volatile substance, in a space permeated by an electric field. This field exerts a force on the drops in a direction opposed to that of gravity. The strength of the field can be so adjusted as to make the drops slowly rise against gravity. On switching off the field the drops slowly fall. In this way Millikan has been able to keep the same drop under observation for several hours. The velocity of the drop, whether rising or falling, is constant, as one would expect on theoretical grounds. But, in the course of a long series of observations, it is sometimes found that a drop changes its velocity suddenly, in a discontinuous manner. Such a phenomenon is exactly what we should expect. We may suppose that one or more free ions have suddenly joined that particular drop. If this be so we should expect the new charge on the drop to be a whole multiple of the elementary charge, a conclusion which is entirely borne out by experiment. As an example we may quote the figures for one particular drop observed by Millikan. It would sometimes gain and sometimes lose, ions, and its successive charges were proportional to the whole numbers, 5, 6, 7, 8, 7, 6, 5, 4, 5, 6, 5, 4, 6, 5, 4. Here is a clear indication that the charge on a drop does not increase or diminish by less than a certain constant minimum charge. We may mention that in determining the actual value of this elementary charge Millikan was not satisfied with Stokes's formula, but used the more complete formula recently worked out by Cunningham.

We have, then, a fairly clear idea of the mechanism of electrical conduction in liquids and gases. In each case we have free atoms carrying a positive or a negative charge, which atoms, under the influence of the electric force, move towards one or other of the electrodes, this double drift constituting the electric current. In each of these cases the electric current really consists of the motion of elementary charges of electricity attached to atoms of matter.

But do these results throw any light on the conduction of electricity in solid conductors? Can we suppose that the electric current in metals is due to the motion of charged atoms of matter? The difficulties in the way of this view seem to be insuperable. In the first place the fact that every conducting circuit is composed of at least two dissimilar substances presents us with only two possible alternatives. Either the moving charged atoms can cross the boundary or they cannot. If they cannot we must get a piling up of material at the boundary. If they can we must find atoms of one substance present in the other substance. Neither of these effects has ever been observed. The most careful examination fails to show the slightest change in any metallic conductor due to the passage of an electric current through it. If the electric current in conductors is due to the movement of particles, therefore, these particles cannot be atoms, for the atoms of different elements have entirely different properties and would reveal their presence in a different substance quite unambiguously. The moving particles, if they exist, must be of a kind common to all elements, or ·at least to all metals, so that a number of them, in a free state, passing round a circuit formed of different elements, leads to no change in the properties of those elements. We shall see that the discovery of electrons gives us exactly the particles required, and clears up the whole question, together with many more important ones, of the conduction of electricity in metals.

It was between 1895 and 1900 that the first discoveries that are fundamental for the modern theory of matter were made, although in some cases these discoveries consisted in the right interpretation of phenomena that had long been known. For instance, certain features that attend the electric discharge through rarefied gases had long been known. Thus Watson, writing in the middle of the eighteenth century, describes the effects observed on passing an electric current through a long glass tube exhausted of air. " It was," he wrote, " a most delightful spectacle, when the room was darkened, to see the electricity in its passage: to be able to observe not, as in the open air, its brushes or pencils of rays an inch or two in length, but here the coruscations were of the whole length of the tube between the plates, that is to say, thirty-two inches." Watson believed electricity to be a fluid and he thought that in this phenomenon he saw the electric fluid " without any preternatural force, pushing itself on through the vacuum by its own elasticity." Nearly a hundred years later Faraday noticed in such a tube that the stream of purplish light stopped short of the negative electrode, leaving a narrow dark space. No further advance worth mentioning was made for twenty years, a fact which, it has been suggested, was probably due to the inefficiency of the air pumps then in use. Certainly the next advance was not made until 1858, three years after Geissler invented his mercurial air-pump. This advance was Plücker's discovery that the electric discharge in a vacuum tube is deflected by a magnet.

The actual phenomena in a vacuum tube are rather complex, but a great part of the effects seems to be due to something streaming from the negative electrode—or *Cathode,* to use Faraday's name for it. Hittorf, Plücker's pupil, proved that this stream travels from the cathode in straight lines, for by interposing an obstacle in the path of the stream he obtained a clearly defined shadow. It was surmised by some that this stream was an ethereal radiation; others, notably Sir William Crooks, supposed that it was constituted of electrically charged particles of matter. These particles he supposed to be ordinary gaseous molecules. There were grave difficulties in the way of both views. J. J. Thomson, in 1894, succeeded in showing that the velocity of the cathode rays was much less than that of light—a fact that makes it difficult to suppose that they are ethereal vibrations. In the following year Perrin proved that the rays actually carry negative electricity with them—which etheric waves could hardly do. On the other hand, Lenard showed that the cathode rays would pass through sheets of metal so thick as to be quite opaque to ordinary light. That particles of matter should be able to pass through such sheets seemed quite impossible.

The essentially correct hypothesis was put forward by

J. J. Thomson in a lecture delivered in 1897, where he says: "From Lenard's experiments on the absorption of the rays outside the tube it follows on the hypothesis that the cathode rays are charged particles moving with high velocities that the size of the carriers must be small compared with the dimensions of ordinary atoms or molecules. The assumption of a state of matter more finely subdivided than the atom of an element is a somewhat startling one; but a hypothesis that would involve somewhat similar consequences, viz., that the so-called elements are compounds of some primordial element—has been put forward from time to time by various chemists." The essentially correct element in this hypothesis is that, in the cathode rays, we are dealing with particles much smaller than atoms. And it was natural, of course, to assume that these particles were little particles of ordinary matter carrying electric charges. Although the "atom of electricity" was a fairly familiar conception, it was always conceived as "attached" to matter.

On the assumption that the cathode rays consist of small electrically charged particles the next step, of course, was to determine the mass and charge of a single particle. This could not be determined directly. What could be determined by direct experiment was the ratio of the charge to the mass. The cathode rays in a vacuum tube can be deviated both by a magnetic and by an electric field. By measuring the two deviations the ratio of the charge to the mass of a single particle can be calculated. The experiment was made by J. J. Thomson, and he obtained for this ratio a value more than a thousand times greater than the value of the same ratio for a hydrogen ion. Now the hydrogen ion is a hydrogen atom carrying the elementary electric charge—and the hydrogen atom is the lightest of all atoms. We may therefore suppose either that the cathode ray particle has the mass of a hydrogen atom but a charge more than a thousand times greater, or we may suppose that its charge is the elementary electric charge and that its mass is more than a thousand times less than that of a hydrogen atom. The latter supposition, even without direct measurements of the charge on a particle, was an irresistible inference. For these cathode particles were found to be identical whatever the nature of the gas through which the discharge passed, and whatever the nature of the metals constituting the electrodes. Also, they could be obtained without any discharge. They are emitted from metals when ultraviolet light is allowed to fall on them. And these particles. so emitted, can be used as nuclei for the condensation of drops as in the experiment of Thomson's we have previously described, and it is found that the charge on a drop is the elementary electric charge. It was concluded therefore, that in these particles we have masses much smaller than that of a hydrogen atom—to be more precise, about 1,830 times smaller. The dimensions, also, of these particles could be calculated, and it was found that a cathode ray particle had a diameter one hundred thousand times smaller than the diameter attributed to the smallest atoms.

And here an exceedingly important question arises, the answer to which revolutionized our notions of matter. We have seen that an electrically charged body owes some of its inertia to its electrical charge. How much of the inertia of a cathode ray particle is to be attributed to its electrical charge? This is a matter for calculation. The answer is that *the whole of the particle's inertia must be attributed to its electric charge.* The particle is *nothing but* its electric charge.

At the time this result was enunciated it was extremely difficult to grasp. We see now that the difficulty came from the persistence of the old idea of matter as substance.

The electric particle, or electron, as it is called, possessed the essential property of matter, namely, its inertia. But the fact that electricity was not regarded as a substance made this notion of " disembodied charges " seem almost paradoxical. Before this notion could be grasped and the electron admitted as a constituent of matter, our conception of matter had to be made more abstract. The notion of *behaviour* had to replace the notion of *substance*. Or the word substance could be interpreted to mean a certain kind and sequence of behaviour. The behaviour of the electron could be summarized by saying that it possessed inertia, that it was of a certain finite size, that it persisted essentially unchanged through time, and that it was capable of motion, that is, of infecting certain successive regions of space with its properties at successive moments of time. These characteristics are sufficient to enable it to be regarded as a possible constituent of what we call matter. To say also that it is wholly composed of electricity is merely a way of summarizing certain other characteristics of its behaviour, and, since these characteristics are not logically incompatible with the others, the electron is not, regarded as theoretical construction, a self-contradictory entity, and may therefore be assumed, if the assumption be found convenient, to exist in Nature. How far this assumption is convenient, in the light of the most modern researches, we shall see later. But it is certain that the theory of matter that has been built up on the electron is very largely true. Before this theory is described, however, we must mention some other discoveries that belong to the same period.

It was towards the end of 1895 that W. C. Röntgen noticed that an unused and protected photographic plate, kept in a room in which vacuum-tube experiments were carried on, showed, on development, distinct markings. This discovery was the result of an accident, but it led

Röntgen to undertake experiments to discover the cause of these markings. He found that a radiation, capable of affecting sensitive plates, emanated from the vacuum-tube when the electric discharge was passing. He also found that this radiation arose at that part of the glass walls of the tube that was struck by the cathode rays. These X-rays, as the discoverer called the radiations, enjoy very remarkable properties. It was shown that they were propagated in straight lines, that they cannot be refracted by any of the substances that refract light, and that they cannot be deviated from their course by a magnetic field. Further, they are able to pass through many substances that are opaque to ordinary and ultra-violet light. Many theories of their nature were immediately put forward. Röntgen's own theory was that they were longitudinal vibrations in the ether. Light, as we know, was supposed to consist of transverse vibrations in the ether. Mathematical theories of the ether showed that it should also be capable of supporting longitudinal vibrations, but no phenomena that could be referred to such vibrations had hitherto been forthcoming. Such phenomena were a sort of " missing link " in ether theory. Röntgen hoped that the missing link had now been found. Other theorists, with different preoccupations, suggested that the famous etheric vortices had now manifested themselves. It was also suggested that the X-rays were infra-red rays, and some thought they were cathode rays that had somehow been " sifted " from the others.

What turned out to be the correct theory was put forward by Schuster a month after Röntgen published his researches. He suggested that the X-rays were essentially like light-waves, but of exceedingly small wave-length. They are not rays carrying electric charges. This hypothesis could not be proved until some way of actually measuring the wave-length of X-rays was invented. The most satisfactory way of doing this would be by passing them through a " diffraction grating," which usually consists of a number of lines ruled very close together on a piece of glass. Such an apparatus " diffracts " waves and enables their lengths to be measured. But in order to do so the interval between adjacent lines must be of the same order of magnitude as the wave-length to be measured. It is sufficiently extraordinary that such gratings can be made for ordinary light. Gratings containing a thousand lines to a millimetre have been ruled. But if, as was suspected, X-rays have a wave-length a thousand times smaller than light-waves, it was obviously hopeless to expect to measure them by a diffraction grating. Human beings could not hope to rule a million lines to a millimetre.

The difficulty was got over by a brilliant suggestion due to Von Laue. Crystals consist of bodies in which the constituent molecules are regularly spaced in regular layers. The spacings of the fundamental intervals are of the order of the molecular diameters, and this was just the order of magnitude attributed to the X-ray wave-lengths. In crystals, therefore, we have our ready-made diffraction gratings for X-rays. These gratings are three-dimensional, it is true, but the influence of that fact on the diffraction patterns was easily taken into account by the mathematicians. The experiment was made, and X-rays were found to be transverse waves, like light-waves, and of the extreme shortness that had been suspected. Amongst the phenomena of a vacuum-tube, therefore, we have these extremely rapid vibrations set up by the impact of the streaming electrons on the walls of the tube. We must now pay attention to yet another phenomenon characteristic of the electric discharge in a vacuum-tube.

The electric discharge in a vacuum-tube, as we have

said, is complex. The actual appearance of the discharge is too complex to allow us to suppose that it is produced wholly by a stream of negatively charged electrons proceeding from the cathode. For some time, however, nothing but cathode rays was discovered in the tube. It was in 1886 that Goldstein, working with a cathode that consisted of a plate perforated by several holes, discovered faint streams of light emerging behind it. These streams excite a phosphorescence where they strike the glass tube, but of a different colour from that produced by the cathode rays. He gave to these streams the name of " canal rays." Since they are proceeding in the direction opposite to that of the negatively charged cathode rays it was natural to suppose that they were positively charged. It was not until 1898, however, that these rays were definitely proved to carry positive charges. Still later experiments cleared up the nature of these rays. J. J. Thomson showed that they are atoms carrying the minimum electric charge. Thus the electric discharge in a vacuum tube generates three things: positively charged particles of atomic mass, negatively charged electrons of much less than atomic mass, and X-rays. We must now consider another phenomenon where the same three things are generated in an entirely spontaneous manner.

Röntgen's discovery of X-rays led Henri Becquerel to undertake certain researches on uranium compounds, in connection with their phosphorescence. He found more than he expected. He found that the metal uranium and all its compounds emit rays that can pass through substances opaque to ordinary light and can affect a photographic plate. Other investigators took the matter up and tried to find these rays in other substances. In 1898 they were found to be given out by thorium and its compounds, but in the same year Pierre Curie and Madame Curie succeeded in isolating a hitherto unknown element which not only gave out these rays, but was a million times more active than uranium. This substance they called *radium*. Certain similarities between the rays given out by these substances and those emitted by a vacuum tube were soon discovered. Thus it was found that these rays, like cathode rays and X-rays, impart conductivity to gases. It soon became apparent, moreover, that the rays emitted by radium are not all of the same kind. Rutherford found, in 1899, that at least two distinct types are present; one type, that he called α-rays, being readily absorbed, while the other type, the β-rays, are much more penetrating. A number of other investigators then showed that part of the radiation is deflected by a magnetic field, and part is not. Finally it was established that the radiation is of three types. There are α-rays, carrying a positive electric charge and having masses of atomic dimensions. These particles are shot out with velocities that may exceed 20,000 kilometres per second. Their penetrating power is nevertheless comparatively feeble. They are stopped completely after penetrating a few centimetres of air. The β-rays are electrons. They are the fastest-moving particles in Nature, sometimes reaching a velocity that exceeds nine-tenths of that of light. They are very penetrating, losing scarcely half their intensity after traversing a metre in air. The third type, the so-called γ-rays, are not particles at all. They may be regarded as very short X-rays and, owing to their smaller wavelength, are more penetrating even than X-rays. They can pass through a centimetre of lead before their intensity is halved.

It early became apparent that all these rays proceed directly from the radium *atom*. No physical or chemical change to which the radium atom can be subjected has the slightest influence on these phenomena. The process consists in an actual disintegration of the radium atom, as was abundantly demonstrated by a number of experiments. The radio-active atom, by shooting out these rays, changes into an atom of a different substance, having a lower atomic weight. The new atom may also be unstable and disintegrate into yet another atom. This process may continue through a long series of changes before a stable atom is reached. Thus uranium, with an atomic weight of 238, passes through a long list of transformations before finally settling down as lead, with an atomic weight of 206. Great light was thrown on these changes by the discovery of the nature of the α-particles. When the ratio of the charge to the mass of an α-particle was determined, it was found to be half the value of that ratio for a hydrogen ion. Two explanations of this fact immediately presented themselves. We might suppose that an α-particle consisted of two hydrogen atoms united with one positive elementary charge. Or we might suppose that we were in the presence of a new element having twice the atomic weight of hydrogen and carrying one unit positive charge. Neither of these explanations was satisfactory, and the true explanation was given by Rutherford. He showed that the α-particle consists of a helium atom carrying two positive unit charges. This makes the ratio right, since the mass of a helium atom is four times that of a hydrogen atom. Rutherford actually managed to collect α-particles in an evacuated space and, on passing an electric discharge through this space, he obtained the spectrum of helium.

The phenomena of radio-activity, therefore, furnish perfectly definite proof that the atoms of radio-active elements (of which a large number are known) are complex structures. Amongst their constituents are electrons and atoms of helium. It may seem strange that we find atoms of helium rather than atoms of hydrogen, but we shall see later what explanation can be given of this fact. The

127

γ-rays we may suppose to be secondary results, produced by the explosive disruption of the atom. The question now arises, are all atoms, and not only radio-active atoms, complex structures? And if so, what are the constituents of these structures, and how are they arranged? The attempt to answer these questions brings us to the electron theory of matter.

CHAPTER VIII

THE ELECTRIC THEORY OF MATTER

W<small>E</small> have seen that there are very good reasons for supposing that atoms have structure, and that the small negatively electrified bodies called electrons enter into this structure. Electrons, so far as we know, are simple bodies. We have no experimental grounds for attributing a complex structure to them. We have seen also that positively charged bodies having the mass of a helium atom enter into the structure of some atoms. But if we wish to construct an electric theory of matter we cannot regard these positively charged bodies as fundamental and irreducible. They obviously cannot enter, for example, into the structure of the hydrogen atom. But certainly positive electricity must enter somehow into the structure of an atom, since the atom is normally electrically neutral. A model of the atom, therefore, must allow for equal quantities of positive and negative electricity. A certain measure of success was achieved by a model put forward by J. J. Thomson. He supposed the atom to consist of a sphere of uniformly distributed positive electricity containing a number of electrons whose total charge should equal that of the positive electricity. It can be shown that the electrons, in order to be stable, will have to arrange themselves in rings. These arrangements have a periodicity very suggestive of the periodic system of the chemical elements. Thus if we start with one electron and add others one by one until we have five, we find that they all arrange themselves in one ring. But a ring of six electrons is not stable. When a sixth electron is added it goes to the centre of the sphere. The presence of this central electron affects the stability of the ring. It can now grow until it holds eight electrons. After this number is reached the next electron added also goes to the centre and also affects, of course, the stability of the ring. In this way we get two rings, an inner and an outer. The final stage of this particular structure is reached when the inner ring contains five electrons and the outer one eleven. The next electron added goes to the centre to start a third ring. And so on. This model represents the periodic properties of the elements very well. But a successful model of the atom has to do a great deal more than that. In particular it has to account for the spectra of the various elements, and here the J. J. Thomson atom failed. It also failed to explain certain very important experimental results which we shall now describe.

When α-particles from radium are passed through matter they undergo a certain amount of dispersion. A pencil of such rays when passed through a piece of metal foil, for instance, is found, on emergence at the other side, to be scattered. Since we suppose the metallic atoms, like all other atoms, to contain electrons, this is an effect that we should expect. For the positively charged α-particles, passing near the negatively charged electrons, will be attracted, and thereby pulled slightly out of their path. This effect, in the course of their journey through the foil, will sometimes be cumulative, and we might anticipate that, in some cases, a fair degree of deflection would be produced. The chances of this can be calculated, with the interesting result that such cumulative effects cannot possibly explain the deflections actually observed. Deflections of 150 degrees have been observed, that is, an almost complete reversal of direction. The swift and massive α-particle must encounter an intense force to be deflected through so large an angle. About one particle in eight thousand is so affected when passing through a sheet of platinum. The comparatively slight disturbing effects of electronic attraction cannot account for this proportion of large deflections.

These were the results that led Rutherford to devise his celebrated planetary model of the atom. In this model we are to imagine the positive charge of the atom concentrated in a " nucleus " at the centre. Round this nucleus circulate electrons sufficient in number to balance the positive charge on the nucleus and thus make the whole atom electrically neutral. If the charge on the nucleus be regarded as extremely concentrated, we have a centre of force sufficient to account for the observed abnormal deflections of the α-particles. We may suppose that, every now and then, an α-particle, instead of passing between atoms or through the cloud of outer electrons, comes practically into contact with the nucleus. The intense repulsive force then exerted swings it round in a hyperbolic path, and it emerges from the metal on the same side that it entered it. This theory lends itself to calculation, and the calculations agree with the experiments. The nucleus is, of course, of subatomic dimensions. The α-particle also, these experiments show, is of subatomic dimensions. Thus although the α-particle has the mass of a helium atom it is vastly smaller than a helium atom. We shall see that it is actually the nucleus of a helium atom, and that in all atoms practically the whole mass is carried by the nucleus.

The actual charge carried by the nucleus of any atom can be determined by these experiments. *Calculation* shows that the deflection undergone by an α-particle depends on the charge on the nucleus of the atom it collides with. *Observation* shows that it depends on the atomic weight of that atom. There is therefore a connection between the charge on the nucleus of an atom and the atomic weight of that atom. For platinum, silver, and copper, the charges on the nuclei were found to be 77·4, 46·3, and 29·3 respectively. The atomic weights of these elements are 195, 107·88, and 63·6. We might say that the atomic weight is very roughly twice the charge on the nucleus. But this correspondence is very rough indeed. When we turn, however, from the atomic weight to the atomic *number,* that is, to the numerical position occupied by an element in the table of atomic weights, we find the correspondence is nearly exact. Thus the atomic numbers for platinum, silver, and copper are 78, 47, and 29 respectively. We may conclude that the charges on the nuclei correspond to these figures exactly, since the discrepancies are within the limits of experimental error. We may, therefore, accepting Rutherford's model of the atom, arrive with him at the fundamental law that the charge on the nucleus of an atom is equal to its atomic number. The *atomic number,* we see, is a very important characteristic of an atom. It is, as we shall see more and more clearly, a much more revealing characteristic than the atomic weight. The two characteristics are obviously not unconnected, but the relation between them is not simple. For the early part of the table the atomic weight is approximately twice the atomic number, but as we go farther down the table this ratio no longer holds.

We now have sufficient materials to describe our atom in some detail. The atom is a sort of miniature solar system. At the centre is the positively charged nucleus. The charge on the nucleus is numerically equal (expressed in terms of the " atom of electricity " as unit) to the atomic number of the atom. Circulating about the nucleus are negative electrons, their number being equal to the atomic number of the atom, since that is equal to the charge on the nucleus, and the electrons (each one of which carries an atom of negative electricity) must exactly balance that charge. Positive and negative atoms of electricity are of the same magnitude, we must remember, although they are of opposite sign. The simplest atom, hydrogen, has 1 for its atomic number. Its nucleus therefore carries a charge of one positive unit of electricity. To balance this we must have one electron circulating round it. We see at once that practically the whole mass of the hydrogen atom must be concentrated in its nucleus. For the circulating electron, as we know, has a mass only one eighteen-hundredth part of that of a hydrogen atom. Therefore the nucleus of a hydrogen atom is about 1,800 times more massive than an electron. If we assume that the whole of its mass is due to its charge this would indicate, as can be shown by calculation, that the hydrogen nucleus is 1,800 times *smaller* than an electron.

The existence of the hydrogen ion helps to confirm our general picture of the hydrogen atom. For, if the hydrogen atom lost its circulating electron it would manifest as a hydrogen atom (so far as mass goes) carrying a unit positive charge. Such atoms are known. But it could not, if our picture is correct, ever manifest as an atom carrying more than one positive charge. And, in fact, no such hydrogen atom has ever been discovered. The next member of the atomic table is helium with atomic number 2. We therefore suppose that its nucleus carries two positive unit charges and that it is surrounded by two circulating electrons. This is confirmed by the fact that we can find helium atoms manifesting one positive charge (having lost one electron) and other helium atoms manifesting two positive charges (having lost both electrons) but never more. How are we to suppose the helium *nucleus,* which carries two positive charges, to be constituted? We are supposing that the hydrogen nucleus is the actual positive atom of electricity. Its charge is equal and opposite to the negative atom—the electron—and its mass is much greater, being practically the mass of a hydrogen atom. Can we suppose that the helium nucleus is composed of two hydrogen nuclei? This would make the nuclear charge right, but it would give a wrong value for the atomic weight of helium. For a helium atom is not twice as heavy as a hydrogen atom, but *four* times as heavy. We therefore suppose that there are four hydrogen nuclei in the helium nucleus and, in order to keep the nuclear charge down to 2, we imagine that these four hydrogen nuclei are combined with two electrons, leaving a resultant charge of two positive units. Thus the actual nucleus of helium is given a rather complex structure. It contains electrons as well as hydrogen nuclei. This complex structure is precisely the α-particle shot out by radium. It is a helium atom that has been stripped of its two circulating electrons.

The example of helium shows us the general method of building up atoms. We construct the nucleus by giving to it a number of hydrogen nuclei equal to its atomic weight. We then add a sufficient number of electrons to these nuclei to bring down the nuclear charge until it equals the atomic number. Having thus constructed the nucleus we imagine a number of electrons circulating round it, the number of these electrons being equal to the charge on the nucleus, that is, to the atomic number. An atom of gold, for example, has atomic weight 197. Its nucleus therefore contains 197 hydrogen nuclei. Its atomic number is 79. We therefore have to add $197 - 79 = 118$ electrons to the nucleus to bring the nuclear charge down to 79. Thus we get the nucleus. And to balance this charge and make the atom neutral we surround the nucleus by 79 circulating electrons.

An objection will here occur to the reader. Many atomic weights are fractional multiples of the weight of hydrogen. An example is chlorine with an atomic weight of 35·456. How can such weights be produced by taking whole numbers of hydrogen nuclei? It would seem to follow, on this theory, that all atomic weights should be whole multiples of the weight of hydrogen. We shall see that this is indeed the case. The difficulty has been cleared up by the discovery that we can have atoms of the same substance but of different weights. Such atoms are called isotopes. The apparently fractional values obtained for so many atomic weights come from the fact that the samples of the substance investigated contain atoms of different weights, and the fractional value is merely the effect of averaging. When such atoms are sorted out it is found, in every case, that the atomic weight is a whole multiple of the weight of hydrogen. The discrepancy introduced by the fact that the weight of hydrogen is taken as 1·008 instead of 1 exactly will be made clear when we come to consider the actual formation of the helium nucleus out of hydrogen nuclei.

Before proceeding further with the development of our atomic model we may consider what confirmation of its main lines we may obtain from that actual disruption of the atom called radio-activity. We have said that radium, on disintegration, passes through a variety of other substances before settling down as lead. There are, in fact, about forty radio-active substances known. Both α and β rays occur, as we know, in the course of their disintegration. Let us consider the effect on an atom of the loss of an α or β particle from its nucleus. An α-particle has atomic weight 4 and carries two positive charges. By losing an α-particle, therefore, an atom will have its atomic weight reduced by four units, and its nuclear charge reduced by two units. Since the nuclear charge of an atom is equal to its atomic number we see that the new atom will occupy a place two steps lower down in the table of the elements. It will now be an entirely different chemical substance, having the properties corresponding to its new place in the table. If the particle lost from the nucleus is a β-particle the atomic weight of the atom will remain practically unchanged. But its charge will be increased by one unit. Hence its chemical properties will alter, for it will occupy a place one step higher up in the table of the elements.

These two kinds of changes can be traced when radium disintegration is examined in detail, although some of the products of radium disintegration have a very fleeting existence, turning almost immediately into something else. Radium, with an atomic weight of 226, loses an α-particle and turns into a substance called radium emanation, with an atomic weight of 222. Radium emanation loses an α-particle and turns into radium-A with an atomic weight of 218. This substance itself loses an α-particle and turns into radium-B, with an atomic weight of 214. At radium-B the process alters. Radium-B loses a β-particle and turns into radium-C. The atomic weight, of course, remains unaltered. At radium-C the chain of transformations acquires two branches. Some radium-C atoms shoot out an α-particle and become radium-C″ with an atomic weight of 210. Radium-C″ shoots out a β-particle, and thus becomes radium-D, also with atomic weight 210. But other radium-C atoms, instead of shooting out an α-particle, shoot out a β-particle, and thus become a substance called radium-C′ with, of course, the same atomic weight of 214. Radium-C′ shoots out an α-particle and thus it also arrives at radium-D with an atomic weight of 210. Thus the two branches meet again. They form a loop on the chain of transformations. Radium-D, by losing a β-particle, becomes radium-E. That in turn loses a β-particle and becomes radium-F, i.e., polonium. Both radium-E and radium-F have, of course, the atomic weight of 210. Arrived at polonium this long series of changes has one more step to take. Polonium, by losing an α-particle, becomes lead with an atomic weight of 206. Having reached lead the process seems to have come to an end. If lead is disintegrating it is at a rate too slow to be observed.

If we start with another radio-active substance, thorium, we again, after a long series of changes, reach lead as a final product. But the lead so obtained differs in its atomic weight from the lead obtained from radium. Its atomic weight is 208. Here we have an illustration of the fact that we can have atoms of the same substance possessing different atomic weights. Our theory of the atom lends itself very easily to the explanation of this state of affairs. We merely have to notice the implications of the statement that the position of an atom in the table of the elements, and therefore its physical and chemical properties, depends wholly upon its nuclear charge. Further, this nuclear charge is provided by hydrogen nuclei, or " protons," as they are sometimes called, combined with electrons. It is obvious that different combinations will produce the same resultant charge. Eleven protons and one electron will give a resultant charge of 10. So will twenty protons and ten electrons. So will an infinity of other combinations. But the atomic weight depends on the actual number of protons present. Thus with eleven protons we should have an atomic weight of 11. With twenty protons we should have an atomic weight of 20. But these atoms, of different atomic weights, would have the same chemical properties.

As a matter of fact, such very wide divergences do not occur, but the variations in atomic weights are sometimes, nevertheless, fairly considerable. Thus krypton has atoms of weights 78, 80, 82, 83, 84, 86. For selenium the masses of the atoms range from 74 to 82. The average weight of the krypton atoms, taking into account the proportions in which these atoms occur in any natural specimen of krypton, is 82·92, and this is the weight found by chemical analysis, since such analysis always deals with samples containing an immense number of atoms. For selenium the average atomic weight is 79·2. Chlorine, whose atomic weight, 35·456, seems to depart as much as possible from a whole number, is made up of two groups of atoms, having atomic weights 35 and 37. These groups, mixed in the proportion of about 3 to 1, give the average weight found on analysing any sample of chlorine. These atomic masses have been measured by the method of electric and magnetic deflections applied to the so-called " canal rays " in a vacuum tube. These canal rays, as we have seen, are positively charged particles of atomic dimensions. They are, in fact, atoms that have lost some or all of their circulating electrons. Atoms of the same substance may suffer different deflections, even when carrying the same electric charge. Therefore their masses must be different. The accuracy that can be obtained by this apparently tricky method is extraordinary. Dr. Aston, who first established the existence of isotopes, claims for his method an accuracy of one part in a thousand.

We have said that one interesting result that follows from the existence of isotopes is that all atomic weights are whole multiples of the weight of the hydrogen atom. This is not quite true. It would be true if the weight of the hydrogen atom was exactly unity. But the weight of the hydrogen atom, in relation to the other elements, is represented, not by 1, but by 1·008. Hydrogen, indeed, is the only atom whose weight is not a whole number. How, then, does it come about that helium, which contains four hydrogen nuclei, has an atomic weight of exactly 4? The fact is a beautiful illustration of one of the great generalizations due to the theory of relativity. As we shall see in more detail later, mass and energy must be regarded as convertible terms. When energy is radiated from a system that system loses a certain amount of mass. It gains mass when energy is added to it. A body when hot is more massive than when it is cold. As a body's velocity increases its kinetic energy increases and therefore its mass increases. In ordinary cases this increase of mass is imperceptible. At the enormous speeds achieved by electrons, however, the difference becomes measurable, and is found to be in accordance with the theoretical calculations.

Now in any ordinary chemical combination that is attended with the development of heat the mass lost through the heat radiated is quite inconsiderable. No experiment could detect that the resultant mass of the compound is less than the sum of the original masses of its constituents. But such combinations are not, after all, extremely stable. A rise in temperature of a few degrees is often sufficient to dissociate a chemical compound. Now we have seen that the hydrogen nucleus is an extremely stable structure. In the tremendous eruptions going on in radio-active atoms it emerges as a perfectly whole unit. The formation of so stable a structure must have been attended by the liberation of a great quantity of energy. The suggestion is that the *mass* of the *energy* liberated in its formation is precisely the difference between the actual atomic weight of helium and four times the mass of the hydrogen atom. Four times the mass of the hydrogen atom is 4·032. The difference between this and 4 exactly is a measure, in terms of mass, of the energy liberated by the formation of the helium nucleus. It is also a measure of the energy that would have to be communicated to the helium nucleus in order to disrupt it. The quantity of energy represented by this figure is enormous. It is sixty-three million times greater than the energy expended in ordinary chemical processes, and this is a measure of how much more stable the helium nucleus is than an ordinary chemical compound. Our most intense source of energy, the α-particles, is insufficient to disrupt such a structure. Even the fastest α-particles provide only a third of the energy required. The nuclei of some other elements, however, have been disrupted. Rutherford has bombarded these elements with α-particles, and succeeded in knocking hydrogen nuclei out of them. The helium nucleus is, therefore, exceptionally stable.

We have seen that the fact that the numerical value of the charge on the nucleus of an atom is the same as its atomic number makes the atomic number a much more important characteristic of an element than is its atomic weight. Another indication of the importance of the atomic number is furnished by certain X-ray phenomena. We have seen that X-rays are generated where the cathode rays in a vacuum-tube strike the walls of the tube. The sudden change of velocity on the part of the cathode rays generates the very small pulses called X-rays. But it is not essential that the rays should strike on the glass of the tube. Any substance may be introduced into the tube and placed in the path of the cathode rays. X-rays are generated in all cases. Now X-rays, like light waves, vary in wave-length. Their wave-lengths can be precisely measured by means of the natural diffraction gratings provided by crystals. The longest X-ray wave-lengths are about thirty times longer than the shortest X-ray wave-lengths. The shorter the wave-length the "harder" the X-ray, that is, the more penetrating.

It was discovered that the X-rays produced by bombarding a substance with cathode rays are of two main kinds. The first kind are due merely to the stoppage of the bombarding electrons. The second kind vary according to the substance being bombarded, and are perfectly definite and characteristic for each substance. Each element, under bombardment, emits X-rays entirely characteristic of that element. These X-rays proceed from the *atoms* of the bombarded substance. If the bombarded substance be a compound, formed of two or more elements, the resulting X-rays are merely a combination of the X-rays that are emitted by the constituent elements separately. With what characteristic of the atom are these X-rays associated? The answer is that as we take atoms of regularly increasing atomic number the corresponding X-ray spectra move regularly in the direction of smaller wave-lengths, that is, the hardness of the X-rays increases in a perfectly uniform manner with the atomic number. So regular is this progression that any gap in the table of the elements is immediately revealed. Thus in Moseley's original research, which established the above law, there was a gap between calcium and titanium. This gap was immediately revealed by the X-ray spectrum leaping a double interval, as it were, in passing from calcium to titanium. The missing element is now known. It is the rare substance Scandium, with atomic number 21. This discovery of characteristic X-ray spectra completely confirms the importance of the atomic number as against the atomic weight. The reader who looks down a table of the elements arranged in order of increasing atomic weights will find three or four places where the order of the elements is inverted. At these places a heavier element is made to precede a lighter one. These inversions were originally decided on by considering the whole complex of physical and chemical properties of the elements concerned. The analysis of the X-ray spectra of these elements entirely justifies these inversions. In arranging the elements in their natural order the atomic weight must be made to yield to the atomic number. Of course, with the discovery of isotopes, the *average* atomic

141

weight of an element becomes of less importance than ever.

The model of the atom that we have hitherto been describing has obviously much to recommend it. The model was devised to account for certain experimental results. It does this successfully, and the further experimental tests suggested by it have also been successful. In the course of establishing this model a very considerable change has taken place in our notions of matter. The properties of matter have been explained in terms of the properties of electric charges, these charges being conceived as occupying small regions of space and as persisting through time. Nevertheless, these charges do not have the definite boundaries we are accustomed to attribute to particles of matter. Each electric charge is the centre of a field of influence that permeates all space. In assigning finite dimensions to the charge we are merely saying that, within these limits, the energy of the field is immensely concentrated. By far the greater proportion of the total energy is contained within the small volumes we attribute to the electric charges. We have to regard matter, not as something separate from the electromagnetic field, but merely as points of condensation of this field. The primary entity, as it were, is the field. Its points of condensation are arranged as in our atomic model. But before we can discuss our atomic model further we have to revise other concepts besides our concept of matter.

The model we have described hitherto was designed, in the first place, to meet the results of experiment. In that lies its strength. The atom certainly behaves as if it had the structure we have postulated. But to the mathematician this structure was quite inexplicable. Such an atom, it could be shown, could not exist. For the rotating outer electrons, which are an essential feature of the model, would continually radiate energy and approach nearer and nearer to the nucleus. Finally they would fall into the nucleus, and the atom, as we have imagined it, would completely disappear. The mathematicians therefore objected to this model that, if it were true, the whole material universe ought to have vanished long ago. This reasoning rested, of course, upon certain premises, and these premises were the classical system of dynamics. This system had justified itself in an immense variety of phenomena and in countless particular cases. It seemed, indeed, to be the most perfectly established branch of physical science. On the other hand, there did not seem to be any substitute for Rutherford's model of the atom which would at the same time meet the experimental results and satisfy the demands of the mathematician. Confronted with such a dilemma scientific men are far too sensible to behave logically. They do not give the whole thing up as an insoluble puzzle. The experimentalists continue to experiment, and the mathematicians begin to question their apparently unquestionable assumptions. It becomes apparent that there are very few indubitable truths that can resist human ingenuity for long. In the present case a solution was reached more easily from the fact that this dilemma did not stand alone. Other phenomena were known that were just as irreconcilable with classical dynamics, and the new concepts necessary to deal with them had already been put forward.

The necessity for these new concepts was first made apparent by certain peculiarities attending the radiation of heat. If we have a number of bodies at the same temperature, enclosed in an envelope opaque to heat, a certain state of equilibrium exists amongst the heat rays radiated by these bodies. The total radiation is, of course, comprised of waves of very different wave-lengths. The same radiation is emitted, at the same temperature, by what is called a " black body," that is, a body that reflects none of the heat it receives. As no perfect black body exists, however, experiments on black-body radiation are performed by using an enclosure such as we have described. A small hole is made in the envelope and the emergent rays of radiant heat are separated out into a spectrum. It is then a matter of experimental evidence to find out how the energy is distributed amongst the various rays. We want to know, for instance, how much of the total heat is carried by the long waves, and how much by the short ones.

The mathematical calculation, based on classical dynamics, shows that as the wave-lengths become shorter the energy contained in them should become greater. This was found to be in flagrant disagreement with the experimental results. Experiment showed that for very long waves, the energy is slight. As the wave-lengths decrease the energy rises at a certain point, to a maximum, and then proceeds to fall towards zero as the waves get still shorter. It was shown that these experimental results were not only at variance with the calculations based on classical principles, but that by no ingenuity could they be reconciled with those principles. An analogy may help to make clear how remarkable these experimental results are. If we start some corks bobbing on the surface of a bowl of water their motion will gradually die down and the water will become slightly warmer. The oscillations of the corks have been transformed into molecular motions of the water. This is in accordance with classical principles. To be analogous to the radiation experiments the corks would have to go on oscillating for ever.

The new concepts necessary for the mathematical description of these phenomena were put forward by Max Planck early in the present century. He showed that a formula correctly representing the distribution of energy in the spectrum could be obtained if we assume that energy is radiated and absorbed, not continuously, but in small finite quantities. Energy is radiated and absorbed by atoms, or *quanta,* as they are called. Not all energy atoms are of the same size. The size of the quantum radiated depends on the frequency of the radiation, being larger the greater the frequency. This hypothesis, although apparently made necessary by the radiation experiments, offers great difficulties. In particular, the whole wave theory of the propagation of energy seems to be irreconcilable with the quantum theory, for the wave theory presupposes a *continuous* distribution of energy over the wave-front, and this supposition appears to be confirmed by various experiments, as we shall see later. The quantum theory, when first put forward, did not actually state that energy was atomic in constitution. It said that, in the radiation experiments, energy acted as if it were. Planck, indeed, had devised a system of " harmonic oscillators " which should give the statistical results actually obtained. But he hesitated to say that the actual atoms of matter behaved in this way; he stated only that their statistical results could be obtained from this machinery. The first man to say that the actual atomic processes themselves obeyed the quantum theory appears to have been Einstein. On this assumption he obtained correct quantitative expressions for another extraordinary class of phenomena.

When light falls on a metal electrons are liberated from the metal. The laws governing this phenomenon are very simple and very extraordinary. The number of the electrons liberated depends on the intensity of the light. The velocity of the electrons liberated depends on the frequency of the light. Thus blue light, which has a shorter wave-length and therefore a higher frequency than red light, liberates electrons having higher velocities than those liberated by red light. It does not affect this result if a very intense red light and a very feeble blue light be employed. At a point where blue light falls, however feeble, an electron is liberated having a greater velocity and therefore a higher kinetic energy, than an electron liberated by red light, however intense. For the much higher frequency of X-rays the paradoxical nature of the phenomenon is revealed very clearly. Suppose we create X-rays by stopping cathode rays of a known velocity. These X-rays are allowed to fall upon a piece of metal. Immediately electrons are projected from the metal having the same velocity and therefore the same energy as the original cathode rays. The remarkable nature of this result may be illustrated by an analogy due to Sir William Bragg. Suppose a plank to be dropped into the sea from a height of a hundred feet. From its point of impact with the sea waves spread out in circles. As these waves become larger they also, of course, become feebler, since the original quantity of energy is spread over a wider front. At a distance of, say, two miles, suppose that one of these waves encounters a ship. On the analogy of the experiments we have been describing we should find that the impact of the wave on the ship tore a plank from its side and projected it one hundred feet into the air.

The paradox of the emission of electrons from metals arises because we have to suppose, from the velocity of the liberated electrons, that the energy of the X-rays arrives at the metal in the form of highly concentrated small bundles, and at the same time we have to suppose, in order to understand the propagation of X-rays and of light generally, that the energy conveyed is uniformly distributed over the surface of a sphere. It seems impossible to dispense with either assumption, and yet these assumptions are contradictory. A sort of intermediate hypothesis has been worked out by Sommerfeld and Debye. They

146

supposed that the X-ray energy was propagated in the form of waves, as demanded by the classical theory, but that the atoms of the metal had the power of waiting until they had accumulated the " quantum " of energy from these waves. The electron was then projected with this quantum of energy. When, however, the waiting time of the atom, the " accumulation time," was calculated, it was found to amount to some years. Hence electrons ought not to be shot out of a metal until X-rays have been falling on it for some years. Needless to say, this was utterly contrary to experience. Experiment shows that the effect is, as near as we can measure, instantaneous.

The quantum theory, as we see, enjoys a very curious position. It belongs to the " abstract " theories in the sense that we cannot imaginatively represent to ourselves the way in which quantum phenomena occur. It is, indeed, entirely opposed to the experiences on which our pictorial faculty has been nourished. We can, for example, imagine a tee-to-tum gradually increasing its speed of revolution. But it is difficult to picture a tee-to-tum jumping from 100 revolutions per second to 101 revolutions per second without ever having a speed between 100 and 101. Yet if energy is essentially atomic that is the kind of thing that must happen. Similarly, the kinetic energy of a falling particle must increase or decrease by jumps. In its flight it must acquire a certain finite number of different velocities, without ever passing through the intermediate velocities. But in all this the quantum theory is not, perhaps, more incomprehensible than action at a distance. That it is " abstract " in this sense is a peculiarity it shares with other scientific theories. The equations of the theory, describing how phenomena actually behave, can be written down, and the fact that we cannot picture what they " mean " is perhaps an irrelevant consideration.

But the quantum theory is not only abstract; it is inadequate. There are certain phenomena that the wave theory can deal with, that the quantum theory has not explained. It appears, moreover, that it simply cannot explain them. It cannot include the wave theory. Further, the wave theory cannot include the quantum theory. Both theories appear to be necessary, and they are contradictory. This is a state of affairs which, in this acute and well-defined manner, has never before existed in science. The attitude of most scientific men, in face of it, is to get on with their investigations and trust to the future. We shall see later what grounds there may be for this trust. In the meantime the contradiction has led some, notably Professor Eddington, to speculate as to whether Nature may not prove finally irrational, that is to say, whether the scientific adventure may not have to be given up. Such doubts are salutary, if only for the purpose of making us realize that science *is* an adventure, and not merely an *inevitable* acquisition of new knowledge.

The indisputable success of the quantum theory in dealing with otherwise inexplicable phenomena led a young Danish physicist, Niels Bohr, to consider whether it could be applied to the Rutherford atom. He found that the theory could be applied if it were somewhat generalized. The process called " quantising " was invented. But it was found that the process must be applied, not to energy, but to a rather more abstract conception called " action," which may be defined as energy multiplied by time. There was no *a priori* reason for this procedure in any given case. The device is adopted in order to obtain formulae that shall agree with experimental results. The assumptions adopted by Bohr may be summarized as follows: the outer electrons of an atom describe orbits about the nucleus without radiating energy. This assumption is in flat disagreement with the classical theory and obviously gets over the difficulty that the rotating electrons, by radiating energy, would ultimately fall into the nucleus. Another assumption is that the electron can rotate round the nucleus only at certain definite distances from it. Intermediate orbits cannot exist. The electron continues to rotate in whichever orbit it happens to be unless it is acted upon by some external force. If an external force does act on it, it passes directly to another possible orbit. During this transition, and only during this transition, the electron radiates energy. This energy is monochromatic, that is, of one definite wave-length. The quantity of energy so emitted is the quantum belonging to radiation of that frequency.

From this theory of Bohr's certain theoretical deductions can be made. They are found to be in beautiful agreement with experiment. This is particularly true of the spectra of hydrogen and of ionized helium. The spectrum of each element, as we have said, gives characteristic lines. These lines correspond to different wave-lengths, and therefore, of course, to different frequencies. It was in 1885 that Balmer discovered that the frequencies belonging to various lines in the spectrum of hydrogen could be represented by a certain very simple formula. This formula was obtained empirically. It was simple but very puzzling. It gave the places of the lines in the hydrogen spectrum, and could be used to predict new ones. But Balmer gave no reason at all for the simple law he had discovered, and no reason seemed to be forthcoming. Bohr's first great triumph was that he gave a complete deduction, from his theory, of Balmer's formula. An essential feature of Balmer's formula for the frequency corresponding to any given line in the spectrum is that there always occurs the difference of two quantities. Bohr showed the meaning of this. The radiation occurs, on his theory, when an electron jumps from one orbit to another. These orbits represent different levels of energy, and it is the difference between their energies which is responsible for the frequency of the radiation sent out by the jumping electron. The formula for the frequency, therefore, is given in terms of this difference. Another spectrum that can be fully accounted for by Bohr's model of the atom is the spectrum of ionized helium, where the helium atom has lost one of its outer electrons. Such an atom is very like a hydrogen atom. The differences are that its nucleus is four times as massive as the hydrogen nucleus and carries twice the electric charge. The influence of these factors can be calculated, and the results are in remarkably exact agreement with experiment.

The hydrogen atom is most stable when the outer electron is rotating in the first orbit—the orbit nearest the nucleus. A jump of the electron between any two orbits is responsible for one definite line in the hydrogen spectrum. Since the spectrum given out by any specimen of hydrogen always shows many lines, we see that, of the immense number of atoms present in the specimen, there are always many where electrons are jumping from one orbit to another. When an electron jumps the reverse way between two orbits it absorbs the same quantum of energy that it radiated. Such jumps are always taking place, but the normal condition of a hydrogen atom is when its electron is describing the first orbit. The diameter of the hydrogen atom in this condition, when calculated by Bohr's theory, is found to be in agreement with the measurements made by quite other methods.

The agreement between theory and calculation that we have described is sufficient to give Bohr's assumptions a very serious scientific standing. But, in the case of the hydrogen atom, a further refinement of calculation, and a further refinement of observation, are of particular interest. The most general orbit that the hydrogen electron can describe about its nucleus is an ellipse, the nucleus being at one focus. These ellipses, according to the quantum theory, can only exist at certain definite distances from the nucleus, and the electron radiates energy by jumping from one ellipse to another. Now it is part of Bohr's theory that these elliptical orbits are described under the inverse square law—the law obeyed by the planets of our solar system in their revolutions round the sun. It is a peculiarity of motion in such an ellipse that its speed is not uniform. A planet, in its journey round the sun, moves fastest in those parts of its elliptical orbit that are nearest the sun. The ellipse is not described at a constant speed. The same applies to the motion of the hydrogen electron. The speed of the electron in its orbit is continually changing. Therefore its energy is continually changing. But we have already seen that relativity theory shows that increase in energy is attended by increase in mass. Therefore the mass of the electron is continually changing. What influence will this variation of mass have on its motion? Sommerfeld has solved this problem and has shown that the effect on the spectrum of hydrogen will be that each hydrogen line will be attended with two or three others very close together. Their separations will be very minute, but they can be calculated. These lines have been observed, and the observed separations agree with those theoretically deduced.

Bohr's theory, as applied to the atoms of hydrogen and ionized helium, is remarkably successful. More complicated atoms do not lend themselves to such thorough treatment. But Bohr succeeded in accounting fairly well for their general properties. Thus he related the periodic properties of the elements to the types of orbits in which the outer electrons of the atoms revolve. These orbits fall into families, the same family recurring for members of the same periodic group. The chemical properties of an atom, besides its visible spectrum, depend on the arrangement of its outer electrons. But although Bohr's theory is the most important that has yet been proposed, it has not yet been able to meet all the demands made on it. Chemists, apparently, find it of very little use. It throws almost no light on the actual details of chemical processes. It cannot even tell us how two atoms combine to form a molecule. The inadequacy of Bohr's atom for chemical purposes has led some chemists to construct atoms of their own. But these atoms give no account of spectra. Further, they lack a satisfactory mathematical basis. For these reasons it seems unlikely that they can ever do more than account for the limited class of phenomena they were invented to account for. With the Bohr atom, on the other hand, we may hope its range of explanation will be extended as its details are further examined. But much work remains to be done. Even in the realm of pure physics the large and confusing region of magnetic phenomena offers difficulties that have not yet been resolved by Bohr's model of the atom. And there are indications, as we shall see, that the whole problem of the atom must be approached with entirely new conceptions.

CHAPTER IX

RELATIVITY

ALTHOUGH, as we have seen, it seemed to be impossible to construct a model of the ether which should give a satisfactory mechanical explanation of electromagnetic phenomena, the general notion of an ether was still regarded as indispensable. Indeed, if electromagnetic actions between bodies are propagated with a finite speed in the form of waves, how can such a medium be dispensed with? And although its properties cannot be worked out in detail there are certain questions about it that can legitimately be asked whatever constitution we suppose it to have. One of these questions concerns the influence of moving matter on the ether. Is the ether everywhere stationary, or does moving matter drag the ether in its neighbourhood along with it? This question, although apparently so simple, provides one of the old and outstanding puzzles of ether theory. Like so many other questions about the ether it received ambiguous or contradictory answers when appeal was made to experiment.

Thomas Young, on the basis of certain astronomical observations, supposed the ether in the neighbourhood of the earth to be stationary. Sir George Stokes, however, tried to show that this supposition was not necessary, and elaborated a theory which permitted the earth to carry along the ether in its neighbourhood. It was subsequently discovered that this investigation was open to serious objections, although Planck showed that Stokes's theory could be saved if we make the rather extraordinary assumptions

153

that the ether is compressible like a gas and also subject to gravity. Lorentz worked out a theory whereby the earth imparted to the ether in its neighbourhood, not the whole of its velocity, but only a fraction of it. Hertz supposed that within matter the ether takes part in the motion of the matter, and that it is also moving in space free from matter.

The contradictions between the theories were paralleled by the contradictions between the facts. Thus Michelson, in 1881, obtained what appeared to be direct proof that the earth carried the ether in its neighbourhood along with it. To understand the experiment we must remember that light travels at the same velocity through space whether it be emitted by a source in motion or a source at rest. This is true of all wave motions, since their speed is wholly conditioned by the medium in which they travel. Thus the velocity of the waves set up by a stone falling in a pond is independent of the velocity of the stone. It depends wholly upon the characteristics of the pond. In Michelson's apparatus two equal arms were placed at right angles to one another, and a ray of light passed along each arm. Each arm was provided with a mirror placed at its far end, and thus each ray was reflected back to the junction of the two arms. One arm was placed in the direction of the earth's motion in its orbit; the other arm, of course, was at right angles to it. If, now, the earth is moving through a stationary ether, a simple calculation shows that the two rays, sent out from the junction of the arms at the same moment, do not get back to their starting point at the same moment. The time taken to swim a mile up stream and down again is not the same as the time taken to swim a mile across stream and back again. Michelson's apparatus was delicate enough to measure the difference in the times taken by the two rays of light, if the ether is stationary. He found no difference. He therefore concluded that the ether, in the neighbourhood of the earth, is carried along by the earth. By repeating the experiment at various heights above the earth's surface, Michelson showed that this influence of the earth on the ether must extend to a considerable distance from the earth's centre, for in no case did he find any difference in the times taken by the two rays of light. From this experiment, therefore, we may conclude that moving matter drags the ether along with it.

Meanwhile an entirely different result was obtained by an experiment performed, in 1892, by Sir Oliver Lodge. He made two rays of light pass round the space between two steel discs. The rays described their closed paths in opposite directions. He then made the discs rotate with great rapidity. If these rotating lumps of matter drag the ether along with them it is obvious that the velocities of the two rays should be affected. No influence whatever on the velocities of the rays could be discovered. We may therefore conclude, from this experiment, that moving matter does not drag the ether along with it.

These experiments, and many others, were designed to answer the question, Is the ether everywhere stationary, or is it carried along by moving matter? By asking this question we get contradictory answers. But we can put our question somewhat differently. Can relative motion between matter and the ether be detected? Here we get a consistent answer, and the answer is No. Michelson's experiment does not reveal relative motion of matter and ether. Neither does Lodge's. If we *assume* a stationary ether, then, of course, Lodge's discs are rotating with respect to it. But his experiment does not furnish us a single effect *due* to this relative motion and therefore revelatory of it. The same may be said of all the other experiments that have been tried. That this is so may be seen by the simple fact that all their results are consistent with the assumption that the ether does not exist at all! No experimental results that have ever been obtained can be interpreted as testifying to the relative motion of matter and ether. If we start by assuming an ether, then some experiments show that it is stationary and some that it is in motion. But if we do not start by assuming an ether, there is nothing in these experiments that makes us assume it.

However, assuming that there was an ether, Lorentz made great advances in electromagnetic theory by assuming that this ether was always and everywhere stationary. He was able, by his theory, to explain effects that seemed, at first sight, at variance with it. The hypothesis of a stationary ether, in Lorentz's hands, triumphed over a wide range of experiments. But there remained the Michelson experiment, which had been repeated, in an improved form, by Michelson and Morley, and still gave a null result. This null result could not be explained away as being on the border line of experimental error. Even one-hundredth part of the expected effect could have been detected by the Michelson-Morley apparatus. Its negative testimony, therefore, was more than a trivial thorn in the side of the stationary ether theory. The situation called for heroic measures, and they were forthcoming. A very brilliant Irishman, Fitzgerald, suggested that all matter experiences a contraction in the direction of its motion through the stationary ether! This notion saved the ether. For, if that arm of Michelson's apparatus that pointed in the direction of the earth's motion was contracted in a certain ratio, the travel times of the two rays would become equal, and therefore the null effect of the experiment would be accounted for. The contraction ratio that would bring this about was easily calculated by Fitzgerald and, as Michelson and Morley had used different substances for the arms of their apparatus, it was stated to apply to all matter.

This hypothesis was purely *ad hoc*. At the time it was enunciated there was no reason whatever to believe it except that it made the Michelson-Morley experiment consistent with the hypothesis of a stationary ether. The contraction necessary is very small. For the earth the contraction in the direction of its motion is only one hundred-millionth of its diameter. This hypothesis was taken up by Lorentz and, although he could give no physical reason for it, he succeeded in showing that it was consistent with the governing equations—the electromagnetic equations. Thus the whole theory of a stationary ether was made self-consistent.

But although the Michelson-Morley experiment was thus accounted for, it was still not perfectly clear that no optical or electrical effects would reveal relative motion between the earth and the ether. The Fitzgerald contraction might produce, for instance, certain optical effects in transparent bodies. No such effects were observed, and a number of electrical experiments that were tried also gave a null result. Now all these effects are electromagnetic effects, for the whole region of radiant energy, optical, electric, or what not, had been reduced to electromagnetism by Maxwell. The theoretical question that lay at the very heart of this problem was, therefore, what laws do electromagnetic phenomena obey for bodies in motion? Maxwell's equations give the electromagnetic laws for bodies at rest. But to decide what modifications must be made in these laws for bodies in motion is a very difficult problem. . Both Maxwell and Hertz had made attempts to solve it, but their solutions were entirely unsatisfactory. Lorentz now investigated this question. The fact that motion between the earth and the ether cannot be detected made him wonder whether the electromagnetic equations are not the same for bodies in motion as for bodies at rest, or rather, led him to find out what we must do in order to make them the same. He found that, for the moving system, we must adopt different measurements of space and of time. For instance, if the moving system be a rod moving in the direction of its length, we must attribute to it a somewhat shorter length than it has when at rest. Also, we must suppose that the time between two events on the moving rod is rather longer than when the rod is at rest. If we do this we shall find that the electromagnetic equations for the moving system are the same as for a system at rest. That is, if we want to keep the electromagnetic equations the same, we must alter the space and time measurements in the way shown by Lorentz. All this was put forward, of course, purely as a mathematical device. But it was certainly a curious and interesting fact that the revised system of length measurement, applied to the moving rod, made it appear contracted in just the ratio that had been suggested by Fitzgerald.

An entirely new turn was given to the whole question by a paper published by Einstein in 1905. He enunciated as a principle, to be justified by its consequences, that none of the electromagnetic phenomena that take place within a system enable us to detect whether the system is at rest or whether it is moving, provided the only sort of motion we consider is motion in a straight line and at a constant speed. It was not asserted that the principle applied to irregular motion in a straight line, or to regular or irregular motion in a curved line. For this reason the principle is known as the " restricted " principle of relativity. The principle does not only deny that electromagnetic phenomena can be used to distinguish between rest and motion in " absolute " space. It also denies that relative motion, provided it be of the kind specified, makes any difference to electromagnetic phenomena. If, for instance, we imagine an aeroplane moving in a straight line and at a constant speed with respect to the surface of the earth, then the principle asserts that no optical or electrical experiments conducted in the aeroplane will differ in their results from similar experiments conducted on the surface of the earth. There would be no way, by comparing the results of these experiments, of discovering that the two bodies were in relative motion. A similar principle had long been known to hold good of *mechanical* phenomena. None of the phenomena governed by Newton's laws of motion can reveal whether the system in which they occur is at rest or in uniform motion in a straight line. In a uniformly moving train the motions of pendulums and of colliding billiard balls pursue their course entirely unaffected by the motion. Newton's laws of motion remain the same, whether formulated for a system in uniform motion or for a system at rest. It is for this reason that, in the attempts to detect the earth's motion through the ether, no appeal was ever made to mechanical phenomena. Einstein's assertion is that not only the laws of mechanics, but also the laws of electromagnetism, remain the same for two systems in uniform relative motion. What we may call the principle of relativity is thus extended to all the phenomena dealt with by physics.

But this is a very extraordinary extension. When the consequences of Einstein's principle come to be analysed it is found that the principle cannot possibly be admitted unless we are prepared to revise our notions of space and time. We can see this most simply if we confine our attention to one only of these consequences. Light, as Maxwell showed, is an electromagnetic phenomenon. If we admit Einstein's principle that all electromagnetic phenomena are the same for systems in uniform relative motion, it can be shown to follow that the *velocity* of light is the same for all these systems. This is only one consequence, but it is a very important and extraordinary one. It seems to be impossible that anything which moves can pass an object going away from it at the same speed with which it passes an object coming towards it. We may agree that an *infinite* velocity could do this, since adding or subtracting a finite quantity to infinity does not alter it. But the velocity of light is not infinite. Light has the perfectly well-known and definite velocity of 186,000 miles per second. This is, it is true, the highest velocity we encounter in the universe, but it is very far from being infinite. We know of electrons that move with more than nine-tenths of this velocity. The principle requires us to believe that, *to an observer mounted on such an electron,* a ray of light would pass the electron with the speed of 186,000 miles per second, whether the electron was moving in the direction of the ray or whether it was moving in the opposite direction. We have said " to an observer," but we do not intend to imply thereby that any merely psychological effect is involved. We may replace the observer by scientific apparatus making the necessary measurements automatically. What is essential is that the apparatus should be mounted on the electron. From the point of view of the electron, as it were, light is passing it at the velocity of 186,000 miles per second. For an observer who does not share the motion of the electron this is not the case, of course. It is quite obvious that something must have happened to the measuring apparatus mounted on the electron.

The assertion that the velocity of light is the same for observers in relative uniform motion is certainly in agreement with experiment. The Michelson-Morley experiment, when performed again at the interval of six months, gave the same value for the velocity of light. But in that time the earth, in its motion in its orbit, was proceeding in the opposite direction to its former one, and this change in velocity is quite considerable. Nevertheless the very delicate Michelson-Morley experiment was quite unable to detect it. Accepting, as an ultimate fact about the universe, that the measured velocity of light is found to be the same by observers in uniform motion with respect to one another, Einstein showed that this fact leads inevitably to the conclusion that their space and time measurements are different. Regarded from a system at rest the lengths of a moving system are contracted, and the clocks of that system are going slow. The alterations increase with the velocity of the system. We have spoken of a " moving " system compared with a system " at rest "; but, as we know of no absolute motion, the moving system may be taken as at rest, and then, of course, the system previously regarded as at rest becomes the moving system. The only difference is that the movement is now regarded as taking place in the opposite direction. Regarded from either system, therefore, the lengths of the other are contracted and its clocks are going slow.

When the equations are formulated that give the precise relations between the measurements adopted by the two systems, they are found to be the same as those put forward by Lorentz as mathematical devices for making the electromagnetic equations the same for systems in motion as for systems at rest. But Lorentz was thinking of rest and motion with respect to the ether—and this might fairly be called absolute rest and motion. But Einstein, by obtaining them for systems in relative motion, shows that the notion of absolute rest and motion is quite superfluous —indeed, irrelevant. Further, he puts forward these alterations in space and time measurements as physical facts, and not merely as mathematical devices. A moving system actually *does* appear contracted to an observer not taking part in the motion. His measurements show the moving system as contracted. To an observer on the moving system nothing has occurred, since his measuring instruments all share in the contraction. His measuring rods have contracted to just the extent, and his clocks have slowed down to just the extent, to enable him to find the figure of 186,000 miles per second for the velocity of light, the same figure found by the observer not taking part in his motion. We can easily see that both the time and space measurements of a moving observer must be affected if he is to get the same velocity for light whatever direction he is travelling in. If the compensation was to be brought about only for light overtaking him it would obviously be sufficient to make his measurements of length contract. He would then find that light, in a given time, had gone a longer distance than it really had (as judged by an observer at rest) and hence he would make its velocity, relative to him, greater than it really was (as judged by an observer at rest). If the contraction was greater the greater his velocity, the ratio being that given by Fitzgerald, the compensation would be perfect. But it is obvious that if, instead of running away from the light, he was advancing to meet it, the fact that he was using contracted measuring rods would be worse than useless. It is the fact that his clocks are also affected that enables the compensation to be perfect in all cases, since to measure a velocity we divide a length by a time.

If, then, we admit that a system moving uniformly with respect to some other system has its standards of length and time altered in a certain ratio, depending on the velocity of the system, then we can preserve the principle that the velocity of light has the same value when measured by observers in uniform motion with respect to one another. Otherwise we cannot preserve the principle. We have said that the alterations in space and time measurements brought about by motion are physical facts. This is true; to an observer not taking part in the motion these alterations have certainly occurred. But it is a physical fact that in the same sense that a building has different appearances from different points of view. No experiments conducted in the moving system would discover the slightest change in the measuring standards. We have to admit, in fact, that length and time-lapse are relative notions, conditioned by the observer's motion, just as the shape of a penny is a relative thing, conditioned by the observer's position in respect to it. If we accept this we must entirely give up the old notions of an absolute space and an absolute time. By " absolute " we mean the same for all observers. If we consider observers in relative motion, they will not form the same estimates of either the distance or the time-lapse between two given events. Even should one observer find the time-lapse between two events to be zero, that is, he judges they occurred simultaneously, this estimate will not be shared by observers who are moving with respect to him. Events which are simultaneous as measured from one system are not simultaneous as measured from another.

As we have presented the argument hitherto we have made the relativity of space and time to spring, as it were, from a determination to keep the velocity of light invariant. The conditions for keeping it invariant have been discovered; we make space and time measurements depend on relative motion. But by doing this we have done more than preserve the velocity of light and the electromagnetic equations unchanged. The formulae connecting the space and time measurements of one observer with those of another (called the Lorentz Transformations since, as we have seen, Lorentz was actually the first to introduce them) have certain mathematical consequences, and some of these can be tested by experiment. One of the most interesting of these deductions is that relating to the mass of a body. Mass, as Newton regarded it, is a completely unalterable characteristic of a body. Whether a body is hot or cold, moving or at rest, its mass remains absolutely the same. Now it is a consequence of Einstein's equations that the mass of a body varies with its velocity. The velocity concerned is, of course, purely relative. Judged by an observer not taking part in its motion a moving body has a greater mass than it has when at rest with respect to that observer. The increase in mass is very slight until we reach velocities approaching that of light. But certain radio-active substances, as we know, shoot out bodies having velocities of this order. For these bodies the increase of mass with velocity is measurable and has been measured. The experimental results entirely confirm Einstein's calculations.

The law according to which the mass of a body increases with its velocity is somewhat peculiar. Thus, if the body is moving with half the speed of light its mass is increased by about one-seventh. At nine-tenths the velocity of light its mass is nearly two and a half times greater than its value at rest. At ninety-nine-hundredths of the velocity of light its mass is seven times greater than at rest. For greater speeds the mass increases very rapidly, and the formula shows that for the velocity of light itself the mass becomes infinite. This can only mean that the velocity of light is a natural limit for moving matter. With respect to any system whatever the motion of a material body can never exceed the velocity of light.

Thus we see that, in this respect also, the velocity of light plays the part of an " infinite " velocity. There is yet another respect, as our equations show us, in which the velocity of light behaves as if it were infinite. According to our ordinary notions of space and time measurements we can superpose velocities on one another according to the ordinary arithmetical law of addition.

Thus we assume that a man walking at four miles an hour along a train that passes a platform at sixty miles an hour, himself passes the platform at sixty-four miles an hour. We assume that this would be his total velocity as measured by an observer on the platform. Einstein's equations show us, however, that the man's velocity relative to the train is not the same for an observer on the train as for an observer on the platform. We should expect this, since the two observers are using different standards of measurement. If the observer on the train makes the man's velocity, relative to the train, four miles per hour, the observer on the platform makes it somewhat less. Hence the observer on the platform will make the man's total velocity somewhat less than sixty-four miles per hour. If we add another velocity on to these we have to go through the same reasoning. If the walking man shoots a bullet down the length of the train the total velocity of the bullet, measured from the platform, is not the sum of its velocity relative to the man, plus his relative to the train, plus that of the train relative to the platform. It is less than this sum. Similarly, the velocity of the bullet relative to the train is not the sum of its velocity relative to the man plus his relative to the train. It is less than this sum. Velocities do not combine according to the arithmetical law of addition. The way in which they do combine is such that the total velocity relative to any system cannot exceed the velocity of light.

If we accept Einstein's theory, therefore, it seems to follow that we live in a universe the space and time of which have such curious metrical relations that the velocity of 186,000 miles per second is an ultimate velocity. On our ordinary notions, where the distance between two events is an absolute quantity (the same for all observers) and where the time between two events is an absolute quantity, the only ultimate velocity is an infinite velocity.

On Einstein's view space and time are not, as it were, the fundamental characteristics they were supposed to be. Each observer has his own space and time in which he places events. But these spaces and times are related. They do not vary at random from one observer to another. We know the precise equations connecting their spaces and times and how these spaces and times depend on their relative uniform motion.

And the fact that there is this connection leads to a very important consequence. Minkowski was the first to bring out clearly that, although different observers inhabit different spaces and times, they all inhabit the same space-time. By space-time he means that four-dimensional continuum which is obtained by assimilating the three dimensions of space to the one dimension of time.

Everything with which science is concerned exists both in space and in time. The full relation between any two events (say two flashes of light) is a spatio-temporal relation. There is a distance between them and also a time-lapse. If we regard time-lapse as a sort of " distance " in the time dimension, we can present the spatio-temporal relation between events as a sort of generalized distance in four dimensions. We can have distance between two points on a line, two points on a plane, and two points in space. These are one, two, and three-dimensional distances respectively. If we add time-lapse, regarded as a distance, to the three-dimensional distance, we get a four-dimensional distance. If each of our observers combines his space and time distances between two events he will get a four-dimensional distance—which is really, of course, a spatio-temporal relation between the two events. Now Minkowski showed that, provided observers combined their space and time distances in a certain way, they would all get the *same* value for the four-dimensional distance between any given

166

pair of events. This four-dimensional distance is called the " interval." Different observers disagree as to the space separating two events, and as to the time separating two events, but they all get the same value for the *interval* separating two events. Thus it is the *four-dimensional distance* of two events which is the *absolute* characteristic of them. It is not space and time taken separately, but space-time, which is absolute.

This way of regarding things, first made clear by Minkowski, gives a remarkable beauty and simplicity to Einstein's ideas. The paradoxes of his theory become quite natural. We see that if the universe is regarded as a four-dimensional continuum, then the spaces and times of different observers are nothing more than different sections, as it were, of this continuum. The interval on whose value all observers are agreed, is split up, owing to their different motions, into different space and time components.

But for this to be true, that is, for the interval to have this absolute quality, it must be constituted in a certain way. The space and time measurements which compose it must be combined in a certain way. They must not be just simply added together, for example. The time measurement must be subtracted from the space measurement. Now this is rather odd, for it shows that the geometry of space-time is not quite the same as Euclid's geometry. We shall discuss non-Euclidean geometries later, but we assume at present that the reader is aware of the fact that Euclid's geometry is not the only possible one. There are any number of others just as possible, and it is a matter of experiment to decide which applies in any particular case. It is a question, at bottom, of the behaviour of measuring instruments in any particular case. Now the essential mathematical difference between different geometries is to be found in the formula they give for the distance between two points. Euclid gives one formula, Lobachevsky gives another, Riemann gives another, and so on. Now the formula for the interval, the four-dimensional distance we have been discussing, shows that the geometry of space-time is not Euclidean. It is very like Euclid's, and for that reason is often referred to as semi-Euclidean. But the difference is important. We can deduce from it that the ultimate velocity in this four-dimensional continuum is not infinite, as it would be if the continuum were Euclidean, but finite. The fact, therefore, that the ultimate velocity of our universe is finite, namely, the velocity of light, becomes a perfectly natural thing if we regard our universe as a four-dimensional continuum possessing a semi-Euclidean geometry.

The philosophical consequences of this discovery are obviously important, but it is difficult to say precisely what they are. It is often said that one of these consequences is that time must be regarded as "unreal." The future must be supposed to "exist" as indubitably as the past. Events do not happen; we come across them. But in saying this it is assumed that the characteristics of time that are found necessary and sufficient for this scientific theory are exhaustive of the nature of time. This is to assume, not only that this theory is permanent, but that everything in our experience of time that can be neglected by this theory is irrelevant to the nature of time. Our experience of an actual creative movement in time is described as "psychological," the word psychological being considered, in some mysterious way, to discredit it. Without going into these questions we may content ourselves by saying that, for those regions of physical phenomena that are covered by relativity theory, a great economy of statement is obtained by locating events within a four-dimensional continuum in which the clear-cut distinction we have hitherto supposed to exist between time and space has been largely obscured.

But whatever relations may ultimately be found to exist between the mathematical concepts of time and space used in physics, and the time and space of consciousness, certain derivatives of the scientific concepts have been altered once for all. A striking example is in the relation the theory shows to exist between mass and energy. We have already seen that a body's mass increases with its velocity. The formula shows that to a first approximation the increase in mass of the body may be regarded as due to the body's energy of motion—its kinetic energy. In virtue of its additional energy it has acquired an additional mass. Mass and energy, it is thus suggested, are convertible terms; they are two names for the same characteristic. Energy, therefore, has weight, and moving energy possesses momentum. It has been experimentally demonstrated that light exerts pressure on any body on which it falls. This pressure is due to the momentum of the moving light energy. It has also been proved, as we shall see, that light has weight. A material particle may be regarded as a very condensed localization of energy. Matter is no longer regarded as permanent in the old sense. A cooling body is losing mass. We have seen how this theory explains the stability of the helium nucleus, and also how the variation of mass with velocity explains the " fine structure " of the hydrogen lines.

CHAPTER X

GEOMETRY AND PHYSICS: THE FINITE UNIVERSE

PERHAPS the most important aspect of the new co-ordination of phenomena given by Einstein's theory is the rôle played by geometry in that theory. This is the aspect that Minkowski was the first to grasp clearly. Certain effects, which, on the old outlook, would have been ascribed to special "laws of Nature," are now regarded as due to the geometrical properties of the space-time continuum. It has been shown, in fact, that geometry and laws of Nature are not independent of one another. As long as it was assumed that all spatial measurements must necessarily conform to one particular geometry—Euclid's geometry—science had only, as it were, one variable to play with, namely, laws of Nature. The mathematical description of phenomena proceeded on the assumption that all measurements necessarily conformed to Euclid's geometry. A triangle formed by straight lines, for instance, must necessarily contain two right angles.

An instance of the way in which laws of Nature are conditioned by such assumptions has been given by Poincaré. Suppose it were found that a triangle formed by a star and by the earth in two positions on its orbit had its interior angles less than two right angles. This would not necessarily be taken to indicate that Euclidean geometry could not be applied to stellar space. Such an assumption would explain the observations, but they can also be explained by assuming different laws of Nature. For it is assumed, in measuring this stellar triangle, that light is propagated in straight lines. If we wish to preserve

Euclidean geometry, therefore, we can do so by assuming a new law of Nature, namely, that light is not propagated in straight lines.

We have two variables in terms of which to describe any phenomenon, geometry and laws of Nature. They are not independent. The choice we make in one region influences the choice we make in the other. The criterion appears to be " simplicity." Thus, in the supposed case of the stellar triangle, retention of Euclidean geometry means a far-reaching alteration in our physics. It might prove to be so extensive an alteration that Euclidean geometry would hardly seem worth the price paid for it. An altered geometry, on the other hand, might lead to all sorts of complications. The advantages and disadvantages of each course would have to be balanced up.

Up to the time that the theory of relativity appeared no geometry other than Euclid's had ever been assumed as the basis of a scientific theory. Indeed, until the invention, about a hundred years ago, of non-Euclidean geometries, the geometry of Euclid was generally regarded as a necessity of thought. The axioms on which Euclid's geometry is based were regarded as unescapable. There was some uneasiness, it is true, about one of the axioms, the axiom concerning parallel lines, but this uneasiness attached chiefly to the clumsy and apparently arbitrary form of the axiom. It was not seriously questioned that the axiom was true. Attempts were made, for something like a thousand years, to deduce this axiom from the other axioms of Euclid. None of these attempts were successful.

Early in the eighteenth century a very able logician, Saccheri, tried the effect of constructing a geometry that denied Euclid's parallel axiom. Saccheri expected, in this way, to be led into self-contradiction but, although he tried very hard, he never succeeded in contradicting himself. What he actually did, although he never realized it, was to construct the first non-Euclidean geometry. The first serious doubts of the validity of Euclid's axiom seem to have occurred to Gauss, but he was afraid to publish his discoveries. The first non-Euclidean geometry to be published was discovered early in the nineteenth century by two men independently, a Russian, Lobachevsky, and a Hungarian, Bolyai. Each of these men constructed a perfectly self-consistent geometry which denied Euclid's parallel axiom. Another non-Euclidean geometry was later constructed by Riemann, and we now know that an infinite number of such geometries is possible.

Geometrical axioms have, in fact, acquired an entirely new logical status. We now know that they are not necessities of thought. What were supposed to be logical necessities are really the outcome of limited experience and deeply rooted habits of mind. We can invent what geometrical axioms we like, provided that they are consistent with one another, and the logical consequences of these axioms form a system of geometry. With this discovery the whole task of giving a mathematical description of Nature assumes a new aspect. We need not assume that physical space, the space in which events happen, is necessarily Euclidean, any more than we need assume that the music of the spheres, should we ever hear it, must be in the diatonic scale. The geometry of space must be determined by experiment. We must observe the behaviour of our measuring apparatus without any *a priori* ideas as to how that apparatus *must* behave. The geometry of actual physical space can only be determined by actual physical apparatus. As long as Euclid's geometry was regarded as a logical necessity it was obvious that if space had any mathematical properties at all they must be consistent with Euclidean geometry. Otherwise the attempt to formulate a mathematical description of Nature would have to be given up. But now that we know that there are an infinite number of geometries all on the same footing, all logically possible, it becomes an open question which of these geometries is most suited to describe natural phenomena. Thus the whole of the Newtonian picture of the world becomes open to examination. Newton assumed, as a necessity of thought, that the geometry of space was Euclidean; that is, he assumed that measurements of spatial relations between bodies, made by rigid measuring rods, would invariably conform to the principles of Euclidean geometry. On the basis of this cardinal assumption he explained the observed motions of bodies in terms of his concepts of mass and force—in particular, the force of gravitation. The question now arises, Can an alternative picture of the world be given if we do not assume that the geometry of space is Euclidean? And if we assume some other geometry for space, what happens to Newton's force of gravitation?

Newton states that the motion of a body acted on by no forces is motion in a straight line with uniform velocity. The " straight line " of which he speaks is the well-known straight line of Euclidean geometry. Such a straight line is defined as the shortest distance between two points. But there are certain assumptions implicit in this definition. On the surface of a sphere, for example, the shortest distance between two points is an arc of a circle, provided we confine ourselves to the surface. Figures drawn on the surface of a sphere do not obey the laws of Euclidean geometry. The interior angles of a spherical triangle, for instance, are not equal to two right angles. If we consider only two-dimensional spaces, that is, surfaces, then it is only for plane surfaces, and surfaces that can be obtained by rolling planes without distorting them (such as the cylinder and cone) that Euclid's rules hold good. Two-dimensional spaces, therefore, can be of different kinds, those where figures obey Euclid's laws and those where they do not. In view of this fact it is rather singular that it was always assumed that three-dimensional space must necessarily be of one kind—Euclidean space. We now know, however, that it is theoretically possible for non-Euclidean three-dimensional spaces to exist. In fact, the number of dimensions makes no difference. We can mathematically design spaces of four, five, or any number of dimensions, having any geometry we like. The " straight line " of a non-Euclidean space is not Euclid's straight line. As we have seen, on the surface of a sphere it is the arc of a circle. Similarly, we can have a three-dimensional space where the straight line is not Euclid's. And the same for a space of any number of dimensions. In each of these spaces there will be a line which is the " straight line " for that space. The general name for all straight lines, Euclidean and non-Euclidean, is *geodesic*. On a plane surface, the geodesic is a Euclidean straight line; on a spherical surface it is an arc of a circle.

We are now in a position to see the rôle played by Euclid's geometry in Newton's picture of the world. Regarding Euclid's geometry as necessarily true of the space in which we are situated, he assumed that the natural unhampered motion of a body in this space was motion in a Euclidean straight line. Any departure from this straight line motion required " forces " to explain it. Since the planets obviously do not move in straight lines they must be acted upon by a force. Thus, on the basis of his assumptions about the geometry of space, Newton was inevitably led to invent his force of gravitation. But it suggests itself that if we did not assume Euclidean geometry for space we may be able to dispense with the force of gravitation. This is the idea that was carried through by Einstein in constructing his generalized principle of relativity. In the first place he was struck by the fact, which we have mentioned in an earlier chapter, that the gravitational mass of a body, and the inertial mass of a body, seemed to be two names for the same characteristic. The most refined experiments do not enable us to distinguish between them. Yet the methods by which they are arrived at seem to be quite independent of one another. The inertial mass of a body is shown by the force necessary to produce a given change in its motion. The gravitational mass of a body is shown by the attractive power it exerts, and which is exerted on it. The two characteristics appear to be quite distinct, and yet they are always rigidly proportional to one another. Is it possible that they are the same thing?

A simple experiment shows that we might attribute phenomena to inertia or to gravitation, indifferently. Let us imagine a man enclosed in a box, and imagine this box to be in a gravitational field—it is suspended, say, just over the Sahara desert. If the man drops a weight it will fall to the floor with an accelerated motion. From this fact he will deduce that he is in a gravitational field. Now imagine that he is transported, with his box, into empty space—that is, to a region far from all gravitating matter. Imagine also that, by means of a rope fixed to the roof, a steady force pulls the man and box upwards. Under the influence of this force the motion of the box steadily accelerates. If, now, the man again drops a weight, it will fall to the floor of the box with an accelerated motion. He would conclude, as before, that he is in a gravitational field. In the first case the weight fell because of its gravitational mass. In the second case it fell because of its inertial mass. But the behaviour of the falling weight is the same in the two cases. In the second experiment we have a case of inertial mass manifesting as gravitational mass. Can it be that this is invariably the case, that when we say that a body "gravitates" it is merely manifesting inertia?

Let us consider what inertia causes a body to do, according to Newton. It causes it to "persist" in its natural state of motion, so that force is required to change that motion. And its natural motion is motion in a straight line—a Euclidean straight line. Suppose now that we accept this property of inertia, but that we also remember that space may not necessarily be Euclidean. We may then generalize Newton's law, and say that the natural, unhampered motion of a body in any particular kind of space is motion in the geodesic of that space. This becomes Newton's law if the space is Euclidean. This gives us a basis for getting rid of the force of gravitation and making all its work be done by some appropriate non-Euclidean geometry. For we may now ask the question, What geometry must space have to make the observed paths of the planets geodesics? If we can find such a geometry, we shall need no force of gravitation. The observed paths of the planets will be their natural paths, which they pursue purely in consequence of their inertia.

This is the problem that Einstein solved, but to solve it he considered, not space alone, but space-time. Einstein, in developing his generalized theory of relativity (published in 1916), fully accepted Minkowski's interpretation of his restricted principle of relativity. But the geometry of the four-dimensional continuum to which Minkowski was led, although not Euclidean, was sufficiently like Euclid's geometry for its straight line to be a Euclidean straight line. The unhampered motion of a body in this continuum was still motion in a Euclidean straight line. It therefore afforded no assistance in the enterprise of abolishing the force of gravitation. Einstein therefore had to replace Minkowski's semi-Euclidean geometry by a geometry much more remote from Euclid's. But he realized, from Minkowski's work, that this geometry was not to be a merely three-dimensional geometry, but a four-dimensional geometry, namely, the geometry of space-time. The geometry was to apply, not merely to the " distance " between two events, but to the " interval " between two events.

If the geometry is to make all the motions attributed to gravitation " natural " motions, that is, motions in geodesics, it is obvious that the geometry must vary from one region of space to another. In the neighbourhood of the sun, for instance, the planets describe ellipses in space. Their four-dimensional path, their path in space-time, is a sort of spiral, and we must remember that it is this path that Einstein was thinking of. But in the neighbourhood of a double star a planet would not describe an ellipse in space, and its four-dimensional path would not be a spiral. The geometry of space, therefore, must be such that it has different geodesics in different parts of space. Space cannot have a uniform geometry. We require, as it were, a very flexible geometry and, as we see from the example of the sun and the double star, the geometry of any region of space must depend on the distribution of matter in that region. The geometry of the space-time region about the sun is not the same as the geometry of the space-time region about a double star. At a great distance from all matter we may assume that the unhampered motion of a body is in a Euclidean straight line, and that the geometry of that region of space-time is the geometry assumed by Minkowski. Since the geometry of any region of space-time varies with the matter present in it, it seems reasonable to suppose that matter acts directly on the space-time continuum, modifying its properties. We can no longer regard the properties

177

of space and time as being entirely independent of matter. On the old view the properties of space and time, as determined by measuring rods and clocks, owed nothing to the presence of matter. On Einstein's theory these properties are influenced by matter. The sun, having a greater mass, exerts a stronger influence than the earth. At the surface of the sun, for example, time runs more slowly than it does on earth, and a vibrating atom, which is a natural clock, vibrates more slowly than it does on earth. Consequently it sends out light vibrations of a lower frequency than the light vibrations sent out by the same sort of atom on earth. This result, which was predicted by the theory, has been confirmed by observation. We are justified in saying, therefore, that matter and space-time are not independent of one another.

As we have presented it hitherto, Einstein's theory would seem to be merely an astoundingly original way of describing phenomena which have already been described. Newton, in terms of Euclidean geometry and the force of gravitation, gave a description of the same region of phenomena. Einstein, by employing a non-Euclidean geometry, has been able to dispense with the notion of a force of gravitation. There is thus a real gain in simplicity. Even if Einstein's theory, compared with Newton's, did no more than cover the same facts in an equally satisfactory manner, its greater simplicity would lead it to be preferred. But Einstein's theory predicts three effects that cannot be accounted for by Newton's theory, and these three effects have been observed.

We have already referred to one of them, that the light sent out by vibrating atoms on the sun is of lower frequency than the light sent out by similar atoms on the earth. The precise numerical value of this effect is given by Einstein's theory, and has been confirmed by observation. The second effect concerns the motion of Mercury's perihelion, mentioned in a previous chapter. According to Einstein's theory even an *isolated* planet would not describe an ellipse round the sun, as it would on Newton's theory, but a slowly rotating ellipse. The amount of this rotation can be calculated from the theory and, in the case of Mercury, is very exactly confirmed by observation. The orbits of the other planets are so very nearly circular that the point of perihelion cannot be so exactly observed, and the effect does not lend itself to such rigorous confirmation. The third effect is concerned with the bending of light on passing near a large mass. Reverting to our image of the man in the box being pulled upwards with a steadily accelerated motion, we see that a ray of light whose direction is at right angles to the man's motion, and which enters through a window in one wall and leaves through a window in the opposite wall, would seem, to the man in the box, to droop slightly towards the floor in its passage through the box. Owing to his own motion across the ray of light, the ray of light would appear to be bent. Since, according to Einstein, a field of inertia is the same thing as a field of gravitation, he predicted that light in a gravitational field, that is, in the space surrounding a mass, would be deflected. During a solar eclipse stars in the line of sight of the sun become visible, and their rays, of course, pass close by the sun. The amount of deflection their light should undergo can be calculated from the theory, and the figures have been verified by measurements made on two eclipse expeditions. A certain deflection would be expected even on the Newtonian theory of gravitation. For energy and mass, as we have seen, are convertible terms. The energy of light, therefore, possesses mass and, like all mass, must be subject to gravitation. Therefore, assuming Newton's force of gravitation, a ray of light passing near the sun should be deflected. But the deflection obtained by this reasoning

179

is only half that predicted by Einstein's theory. The observed deflections are those predicted by Einstein, and not those that can be calculated on the Newtonian theory.

These experimental results are sufficiently remarkable to show that Einstein's theory, as an instrument for the mathematical description of phenomena, is to be preferred to Newton's. These are the only instances where Einstein's theory leads to results which are observationally different from those obtained on Newton's theory; and the general run of dynamical phenomena, which witness to the truth of Newton's theory, witness equally to the truth of Einstein's theory. And where the results diverge, the advantage, as we have seen, lies with Einstein. But besides these empirical confirmations, Einstein's theory greatly commends itself as complying with certain powerful philosophic requirements. In his restricted theory of relativity Einstein was concerned to obtain expressions for the laws of Nature which should remain the same for systems in uniform relative motion. He assumed that the true relations between phenomena could not be dependent upon such a local and accidental peculiarity as the observer's motion, any more than we can assume that the law of some natural occurrence is modified according as to whether the occurrence takes place on an observer's right or on his left. Laws which make explicit reference to such conditions of the observer cannot be true laws. This assumption is, we admit, an assumption, but it is an exceedingly attractive one. Once we admit it, however, we are led to wonder why the laws of Nature should be invariant only with respect to systems in uniform relative motion. Why should not the laws of Nature be independent of the observer's motion, *whatever* his motion?

It is therefore a striking merit of Einstein's theory that.

he succeeds in expressing the laws of Nature in a form which is the same for all observers, whatever their motions, and whatever their systems of measurement. Einstein's theory enables us to isolate those absolute features of the world which are entirely independent of the observer. For this reason Einstein's theory of relativity could justly be called the theory of absolutes, and, if it had been so called, many popular misunderstandings of it would have been avoided. Nevertheless, it is true that most of the " absolutes " of pre-relativity physics have been shown to be relative. "Force " is not an absolute feature of the world. Neither is potential energy. Mass, as we have seen, may be regarded as energy, but even the conservation of mass or energy and momentum is not strictly true. If, however, the four-dimensional space-time continuum be split up into space and time in a particular way, then the conservation of mass (or energy) and momentum holds good for observers using that particular space and time. It is highly significant, therefore, that this particular space and time is precisely the space and time of our perceptions. It is as if the human mind had divided space-time up in the particular way it has in order to find a *permanent* world.

Einstein's great achievement, in his generalized theory of relativity, has been to reduce gravitational force to geometry. In doing so he found that Newton's law did not quite accurately represent gravitational effects, a conclusion borne out by observation. The new equations, which represent the gravitational effects correctly, are really equations concerning the geometry of the space-time continuum. The geometry assumed by Einstein for this purpose is a very general geometry first studied by Riemann. Just as we call a spherical surface, in contrast to a plane surface, "curved," so we may call a space-time continuum governed by Riemann's geometry, a

" curved " continuum. But such a space-time need not be uniformly curved. The geometry is so general that it admits of different degrees of curvature in different parts of space-time. It is to this curvature that gravitational effects are due. The curvature of space-time is most prominent, therefore, around large masses, for here the gravitational effects are most marked. If we take matter as fundamental, we may say that it is the presence of matter that causes the curvature of space-time. But there is a different school of thought that regards matter as due to the curvature of space-time. That is, we assume as fundamental a space-time continuum which is differently curved in different parts of it. These local peculiarities of the space-time continuum manifest to our senses as what we call matter. Both points of view have strong arguments to recommend them, as we shall see later. But, whether or not matter may be derived from the geometrical peculiarities of the space-time continuum, we may take it as an established scientific fact that gravitation has been so derived. This is obviously a very great achievement, but it leaves quite untouched another great class of phenomena, namely, electromagnetic phenomena. In this space-time continuum of Einstein's the electromagnetic forces appear as entirely alien. Gravitation has been absorbed, as it were, into Riemannian geometry, and the notion of force, so far as gravitational phenomena are concerned, has been abolished. But the electromagnetic forces still flourish undisturbed. There is no hint that they are manifestations of the geometrical peculiarities of the space-time continuum. And it can be shown to be impossible to relate them to anything in Riemann's geometry. Gravitation can be shown to correspond to certain geometrical peculiarities of a Riemannian space-time. But the electromagnetic forces lie completely outside this scheme.

This was the state of affairs that led Hermann Weyl to invent his extension of Riemann's geometry. He found that it was possible to construct an even more general geometry than Riemann's, with the result that new geometrical quantities emerged besides those that had been identified with gravitation by Einstein. And the mathematical relations between these new quantities were found to be precisely the same as Maxwell's equations between the electric and magnetic forces. It was natural, therefore, to identify these geometrical quantities with the electric and magnetic forces, to say that the electromagnetic forces were merely the way in which these geometrical quantities manifested themselves to us, just as the curvature of space-time in the region of matter manifests itself to us as a gravitational force.

But this achievement differed profoundly from that of Einstein. A geometry may be defined as a description of the behaviour of measuring appliances. Einstein's geometry describes the actual behaviour of measuring appliances (vibrating atoms, rays of light, etc.) in actual physical space-time. But Weyl's geometry does not describe the actual behaviour of measuring appliances. It is a logically possible extension of Riemannian geometry—there *could* be such a space-time. But there is no evidence, in our world, for the behaviour it attributes to measuring appliances. For this reason it is now regarded merely as a graphical representation of phenomena. It gives us valuable information about the connections between phenomena in the world, just as a temperature-chart gives us valuable information about a patient. But while Einstein showed that the geometrical peculiarities of his space-time *must* manifest themselves to us as gravitational force, Weyl does not show that his additional peculiarities *must* manifest themselves to us as electromagnetic forces. It is certainly remarkable and interesting that he has found a geometrical interpretation of them, but in doing so he has not shown that they have their actual geometrical correlatives in our space-time.

Even if Weyl's geometry be no more than a graphical representation, however, it cannot be dismissed as valueless. For from the relations between the elements of a graph we can deduce relations between the entities symbolized in the graph. This method has been pursued further by Eddington, who has constructed an even more general geometry than that of Weyl. These different geometries, advancing from Euclid's to Eddington's, become, as it were, more and more amorphous. Eddington postulates for his four-dimensional continuum almost the minimum of structure conceivable. Compared with the firm outlines of Euclidean space it is almost a chaos. And yet from this almost structureless continuum he manages to derive, by mathematical analysis, quantities which behave in just the same way as the quantities of our physical perceptions. The equations of mechanics, gravitation, electromagnetism, are found to be obeyed by quantities which have been deduced as pertaining to the metrical characteristics of this continuum. It would appear, therefore, that these laws of Nature throw almost no light upon the actual structure of the world. Provided that the space-time continuum has *some* structure, the universe we know must arise. But more than the universe we know can be deduced from the primitive space-time continuum. Other aspects that may be mathematically deduced do not enter into our experience of the universe. Eddington suggests that this fact indicates that the universe we know is the product, not only of the raw material of space-time, but also of the mind's selective action on this raw material. The mind, in its search for permanence, isolates certain features characteristic of the raw material and pays attention only to those. As a consequence of this selective action the laws of mechanics, gravitation, etc. arise.

This theory is extremely interesting and covers a great deal of ground, but it by no means gives a complete account of the observed behaviour of matter. It does not account for the existence of electrons, nor for the existence of two sorts of electricity, the positive and the negative. Further, it does not throw the faintest light on quantum phenomena. Eddington's derivation of the material universe from an almost structureless four-dimensional continuum must be regarded, therefore, as having achieved a significant but limited success. In any case, the precise status of these new geometries, considered as physical theories, is rather difficult to determine. If they are to be considered purely as graphical representations of phenomena, it is not easy to see what conclusions may be drawn from them. Their philosophical implications are doubtless valuable, but their chief value in physics appears to be that they are suggestive of certain relations that would not otherwise be suspected. But it does not appear to follow that a mathematical necessity in the graph must be an actual law in the physical world. In this these geometries are on quite a different footing from the geometry adopted by Einstein. It is interesting to note, however, that Einstein, who was at first somewhat out of sympathy with these developments, has since made certain investigations where he uses them as a basis. The precise meaning of these developments, however, must at present be left in doubt. Meanwhile, the geometrical interpretation of Nature, which we have traced from Minkowski to Eddington, stupendous as its achievements have been, leaves the actual constitution of matter an almost complete mystery. In the words of Hermann Weyl: " The laws of the metrical field deal less with reality itself than with the shadow-like extended medium that serves as a link between material things, and with the formal constitution of this medium that gives it the power of transmitting effects."

In the general theory of relativity the age-old attempt to give a mathematical description of natural phenomena has achieved its greatest triumph. But in doing so it has made very clear the conventional element in scientific theories. The concepts in terms of which Newton gave his description of phenomena have been replaced by a different set. That mere fact shows that his concepts were not *necessary* concepts. Yet it cannot be doubted that to Newton such terms as mass, force, etc., were names given to actual constituents of the universe. They were not merely convenient terms in a convenient scheme for representing natural behaviour mathematically. They were as definitely names for objectively existing entities as were the names Adam gave to the animals in the Garden of Eden. We see now, however, that these terms are not names of entities, but are concise descriptions of behaviour. They tell us nothing about *what* is behaving. This is true of all the terms used in physics. To say that matter is composed of protons and electrons, for example, does not tell us what matter *is,* but how it behaves. Similarly, Einstein's Riemannian four-dimensional continuum is a description of the behaviour of phenomena—these phenomena being, in the last resort, our perceptions. We can, if we prefer it, use a different scheme. We may, for instance, use Euclidean geometry and supplement it by a host of special " forces." But we prefer Einstein's theory simply because these alternative schemes are more ugly and more complicated.

Science tells us nothing about the world except its mathematical structure. There are various ways of presenting this structure. One way may be in terms of geometrical concepts; another way may be in terms of such concepts as mass, force, etc. All these presentations are logically on the same footing, provided they issue in equally accurate descriptions of the observed behaviour, that is, correlate our perceptions in an equally accurate manner. But although these presentations may all be equally accurate, they are not all equally satisfactory. That is because we have aesthetic preferences. There has been, it is true, an attempt to introduce a criterion of scientific theories which is not obviously merely aesthetic. That criterion is that none but observable phenomena, or phenomena definable in terms of physical processes, shall be imported into scientific descriptions. On this basis Einstein objects to Newton's invocation of absolute space as the cause of the centrifugal forces developed by a rotating body. Absolute space, says Einstein, cannot be observed. It is doubtful, however, whether this criterion can be consistently carried through, even in relativity theory, and it remains true that, on the whole, our selection amongst equally accurate theories is dictated by the criterion of simplicity—an aesthetic criterion. Simplicity, by itself, is not, of course, enough. Thus, to say that everything happens as it does happen by the will of God is not a scientific theory, simple or otherwise, for the will of God is incalculable.

The theory of relativity is to be regarded as the most profound and comprehensive scheme that has yet been proposed for making orderly and coherent those aspects of Nature that can be treated mathematically. Its great merit lies in the fewness of its assumptions and in the immense coherence it bestows upon the world of physics. A still greater unification is bestowed upon the world of physics if we adopt the geometries of Weyl and Eddington, but, as we have said, it is rather difficult to know what precise significance is to be attached to these geometries. We may at present assume, therefore, that the Riemannian geometry of Einstein describes the actual metrical properties of space-time.

But now an interesting question suggests itself. How does space-time come to have these metrical properties? Are we to suppose that space-time possesses an intrinsic structure, modified by the presence of matter, or are we to suppose that the whole of the structure of space-time is due to matter, that space-time without matter would be completely amorphous? Indeed, can we even assume that space-time could exist at all without matter? These questions are all highly speculative, and different answers are given by different writers. Eddington, for example, starts with space-time as the fundamental existent, and accounts for matter as the way in which our minds perceive certain aspects of its structure. Einstein, on the other hand, makes the existence of space-time dependent on matter. Without matter the whole universe would shrink to a point. Such reasons as there are for believing one rather than another of these answers are connected with the investigations that have been made to determine whether the universe is finite or infinite. In pre-relativity physics, of course, this question could not arise. As long as the geometry of space was supposed to be Euclidean it was inconceivable that it should be finite. For it could not be finite without possessing boundaries, and the notion that space is bounded is obviously absurd. Space, therefore, was necessarily infinite.

But the infinity of space brought with it certain difficulties relating to the distribution of the stars. We may assume either that the stars are scattered more or less uniformly throughout infinite space, or we may assume that they are contained within a finite volume, so that the whole stellar universe forms an island in infinite space. Neither supposition is very satisfactory. If the stars are scattered throughout infinite space it can be shown that the night sky should be much more luminous than it is. This difficulty can be overcome, however, if we assume that the percentage of dark stars increases in the remoter regions of space, or if we assume that light is gradually absorbed in its passage through space. Another objection is that such a distribution of stars could not persist if we assume Newton's law of gravitation. It can be shown that the stars would, in time, concentrate into a nucleus. The other supposition, that the stars exist as an island in infinite space, is almost equally unhappy. Under the influence of their mutual attractions the stars will be moving with certain velocities. The velocities of different stars will, in general, be different. These velocities will form, as it were, a certain total range of velocities, and it can be shown that some of the stars should be moving with immense speeds. This is the calculated result of such a distribution of stars, and it is completely at variance with observation. Observation shows that the star velocities are extremely low. This difficulty, which arises out of Newton's law, is not diminished when we substitute Einstein's law.

It was this fact that led Einstein to propose his theory that the universe is finite but unbounded. The distinction between "infinite" and "unbounded," which does not exist for a Euclidean space, is a perfectly real distinction for a non-Euclidean space. Thus the surface of a sphere is an example of a two-dimensional continuum which is finite but unbounded. A body, or a ray of light, moving on the surface of the sphere in any direction, would nowhere meet with a barrier to its further progress. It could move for ever without ever leaving the surface of the sphere. Nevertheless, the total area of the spherical

189

surface—the total amount of " space "—is finite. Similarly, it is mathematically possible for a three- or four-dimensional continuum to possess such metrical properties that it nowhere exhibits a boundary and yet has a total volume which is finite. It is of little advantage for us to try to *picture* such a continuum. We must be content with the logical consequences of certain premises. If the laws governing the behaviour of our measuring appliances —which may be freely moving bodies and rays of light —are of a certain kind, then we can deduce, mathematically, that the continuum having those metrical properties is finite and unbounded. A crude estimate of the size of the finite universe propounded by Einstein shows that a ray of light would go all round it in about a thousand million years. On its journey, however, the ray would be somewhat deflected by the gravitational fields through which it passed and would probably also suffer a certain amount of absorption. For these reasons it is unlikely that it would return, accurately focused, to its starting-point. Otherwise we might suppose that some of the stars are really " ghosts " at the places where stars used to be.

Einstein's finite universe is such that its radius is dependent upon the amount of matter in it. Were more matter to be created, the volume of the universe would increase. Were matter to be annihilated, the volume of space would decrease. Without matter space would not exist. Thus the mere existence of space, besides its metrical properties, depends upon the existence of matter. With this conception it becomes possible to regard all motion, including rotation, as purely relative. The centrifugal force developed by a rotating body, which Newton referred to absolute space, may now be referred to the presence of the other masses in the universe. A body's inertial mass becomes wholly conditioned by the presence of other bodies in the universe. An isolated rotating body would develop no centrifugal forces. In the opinion of some writers Einstein's finite universe gives altogether too important a rôle to matter. Matter is taken as fundamental, and everything else as derivative from it. Eddington's view, as we have seen, is very different. He starts with a four-dimensional continuum, possessing a certain degree of structure. He then makes matter to be nothing but an exhibition of certain peculiarities of this structure. The fact that matter has an atomic constitution, however, has not yet been explained by this theory. In fact, the whole theory of relativity has thrown almost no light upon the problem presented by the atomic constitution of matter.

A finite universe, having a somewhat different structure from Einstein's, has been developed by De Sitter. This universe illustrates very well how widely the universe of science may differ from the universe of common sense. Thus, in De Sitter's universe, natural processes occurring at a distance from the observer will seem to be taking place more slowly than similar processes in his neighbourhood. This slowing down of all natural processes takes place in such a way that at a certain distance—the " horizon "—everything would appear to be at a standstill. This phenomenon is illusory in the sense that if the observer were transported to the horizon he would find everything in his new position proceeding normally, and the horizon would now appear to be at the place he had left. Some evidence for this theory is provided by the spiral nebulae, the most distant objects in the universe. The spectra of a curiously high percentage of these bodies show a displacement towards the red end of the spectrum. This has usually been taken to indicate a high velocity of recession on the part of these bodies. It may be, however, that this displacement is partially due to the general slowing down of all processes, including atomic vibrations, becoming perceptible at these great distances. Unlike Einstein's finite universe, De Sitter's space-time continuum does not appear to be dependent upon the existence of matter. It has an inherent structure of its own.

CHAPTER XI

NEW PROBLEMS

WE have seen that the attempt to describe the intimate structure of matter mathematically led to the Rutherford-Bohr model of the atom. This model achieved, as we have said, some striking successes. By the year 1925, however, it became apparent that certain experimental results could not be accounted for by this model. Calculation could not give these experimentally observed data, not because the calculations were too difficult, but because the Bohr atom did not furnish the necessary premises. The Bohr atom, it could be shown, was too simple to explain certain spectral effects. It did not permit, within itself, a sufficient degree of variety. It became necessary, therefore, to investigate the assumptions made in constructing the Bohr atom, in order to determine whether any of them implied restrictions which could conceivably be loosened. As a result of this examination it was suggested that the notion of the " electron," as assumed by Bohr, could be further complicated. The only motion that Bohr had attributed to the electron was a motion of translation. The electron could move in straight or curved paths, but that was the only kind of motion it was capable of. But the electron was not a point; it was a body of finite dimensions. It was conceivable, therefore, that it could execute a *spinning* motion about an axis passing through it. This notion of the spinning electron introduces an additional complication into the Bohr atom, and gives it a greater degree of variety. The additional variety so introduced was found to be just what is required to explain certain experimental results.

Independent of this development, however, an entirely new attack, which is still being vigorously pursued, has been launched upon the whole problem. We have previously said that a methodological rule which is becoming of more and more importance in physics is that none but observable factors shall be invoked in the construction of a scientific theory. Now the Bohr atom only very partially obeys this criterion. The only *observable* atomic quantities are those characteristic of its radiations. On Bohr's theory the atom only radiates when it jumps from one state to another. It is the passage of electrons from one orbit to another within the atom that is responsible for the observable radiation effects. But Bohr's theory professes to tell us what is happening in the meantime, when the atom is in a "steady" state. The electrons are then revolving steadily in their orbits without executing jumps. Thus Bohr's description of the atom includes an account of atomic states that can never be observed. Another unsatisfactory feature of Bohr's atom is that it contains within itself no prophecy of its future. If we are given a complete specification of the Bohr atom at any instant, we cannot deduce from that specification what the atom is going to do next. We do not know when an electron will jump or where it will jump to. In this respect Bohr's atom is very different from the dynamical systems to which science is accustomed. From the complete specification of the solar system at any instant, for example, its future history can be predicted—which is why eclipses can be calculated. But from a knowledge of the present state of affairs regarding a Bohr atom we cannot say whether or not it will proceed to radiate energy nor, if it radiates energy, can we say how much it will radiate. The Bohr specification seems to make the electronic movements products of electronic freewill, and we cannot but regard such an appearance as illusory. But the reason for this inadequacy of the theory may lie very deep, in the scientific conception of time as mere succession. However, a very promising way of attacking the whole problem, and one that obeys the criterion respecting "observable factors" has been inaugurated by Heisenberg.

Heisenberg gives a specification of the atom only in terms of observable quantities, such as the frequencies of the radiations emitted by it. We have seen that, on Bohr's theory, radiations are emitted only when the atom passes from one state to another. The mathematical expression for the frequency of the radiation therefore naturally involves a reference to both states. A sort of ghostly counterpart of this is to be found in Heisenberg's representation. A frequency is expressed as the difference between two terms. But, in this very abstract representation, we must not simply conclude that the two terms refer, one to an earlier, and one to a later, state of atom. In fact, the physical meaning of the representation is, at present, by no means clear, as Heisenberg admits. He speaks of "the fact that in the theory hitherto the question of the temporal course of a process has no immediate meaning, and that the concept of earlier and later can hardly be defined exactly." Heisenberg does not attribute to the electron the degree of reality that we are accustomed to attribute to objects in the external world. The electron is not, for example, an enduring something that can be tracked through time. Its mathematical description does not involve that degree of definiteness. Any picture we form of the atom errs, as it were, by excess of solidity. The mathematical symbols refer to entities more indefinite than our pictorial imagination, limited as it is by experience of "gross matter," can construct.

Our pictorial faculty seems to be limited by the fact that there are certain properties, attributable to an entity, which we cannot imaginatively dissociate. And yet the dissociation may be quite possible logically. In the development of modern science the pictorial faculty is proving more and more of a hindrance. Thus when Heisenberg tells us that to an electron may be assigned a precise position or a precise velocity, but not both, we must admit that he is talking about an entity we cannot possibly picture. If, however, this entity, regarded as a logical construction, is self-consistent, the fact that we cannot picture it is surely irrelevant. If it be proved that such an entity is necessary in order to co-ordinate our observations, it must be accepted together with its implications. It is quite possible, in fact, that the universe will prove to be a much more elusive affair than we have hitherto supposed. The entities dealt with by science are abstractions, it is true, but there are abstractions that we can " solidify " by connecting them with elements of our experience. Thus the " atom " of Newton's time can be pictured as analogous to a very tiny grain of sand. Also an electron, once we realize that electricity possesses inertia, can be pictured in something the same way. But the electron of Heisenberg's theory cannot be visualized in this way, and perhaps cannot be visualized at all.

The present tendency of physics is towards describing the universe in terms of mathematical relations between unimaginable entities. It would be interesting to speculate as to how far this tendency is due to a particular kind of mind being in the ascendant. There are, and always have been, many scientific men who are resolute visualizers, and who entirely sympathize with the " plain man's " conception of " explanation." It would seem that, at the present time, such men are either experimentalists, or are engaged in branches of mathematical physics which can still be pursued independently of the new conceptions. The history of science, even as it has been sketched in this volume, makes us suspect that science is a fairly flexible construction. One very important conditioning factor of scientific theories is the contemporary state of scientific instruments. A scientific theory has reference to observations, and these observations are conditioned by the instruments used in making them. But there is no convincing reason to suppose that a given set of observations uniquely determines a scientific theory. We have to allow, also, for the type of mind which, as an historical accident, attained the level of genius at that moment. Nevertheless, although something must be allowed for the subjective factor, we cannot suppose that a given set of observations lend themselves to an indefinite variety of interpretations. It is possible that science can no longer be pursued except in terms of the sort of abstractions that are now being used.

In this connection it is interesting to notice that an alternative theory to Heisenberg's, one due chiefly to Schrödinger, and which issues in the same equations, seems, at first, to be more easily picturable. Just as the theory of light as " rays " has to be replaced, for the description of certain phenomena, by the theory of light, as " waves," so the conception of " masses " moving in " orbits " is replaced, in Schrödinger's theory, by trains of waves. A hydrogen atom, on this theory, is a region permeated by waves. The waves fall off very rapidly and become inappreciable at a distance from the centre, which is found to be the same as the empirically determined radius of a hydrogen atom. But when we enquire into the physical meaning of the symbol that obeys these wave equations we find that it has no direct physical meaning. The pictorial imagination, which seizes on waves as intelligible, is baffled by the fact that the quantity which is waving has no direct physical significance. We are again in the region of logical constructions which are not picturable.

In spite of the immense degree of co-ordination effected by relativity theory the science of physics, at the present time, is very far from being a unity. The scientific ideal of giving a mathematical description of all natural phenomena in terms of a few simple entities and principles seems farther from realization than it has been at any period since Newton. Our greater knowledge has given us a sight of deeper difficulties and more irreconcilable facts. And yet the impression is strong that we are on the eve of some great illumination, as if the physics of the moment is in the darkness that precedes the dawn. At present special methods are devised for special problems. These special methods are not obviously connected, and yet, if Nature is a unity, we must suppose the problems to be connected. It is possible that these special methods will turn out to be partial aspects of some great generalization, and that the difficulties they now present, their enigmatic quality, are due to their partial character.

It is probable that the most far-reaching changes in our concepts will occur in connection with the quantum theory, and its relation to the wave theory of light. Two passages, admirably illustrating the disadvantages and advantages of the quantum theory, may be quoted from Sir J. H. Jeans and Dr. Ellis respectively [1]:

" If, however, radiation is to be compared to rifle bullets, we know both the number and size of these bullets. We know, for instance, how much energy there is in a cubic centimetre of bright sunlight, and if this energy is the aggregate of the energies of individual quanta, we know the energy of each quantum (since we know the frequency of the light) and so can calculate the number of quanta

[1] Quoted by Bertrand Russell: *The Analysis of Matter.*

in the cubic centimetre. The number is found to be about ten millions. By a similar calculation it is found that the light from a sixth magnitude star comprises only about one quantum per cubic metre, and the light from a sixteenth magnitude star, only about one quantum per ten thousand cubic metres. Thus, if light travels in indivisible quanta like bullets, the quanta from a sixteenth magnitude star can only enter a terrestrial telescope at comparatively rare intervals, and it will be exceedingly rare for two or more quanta to be inside the telescope at the same time. A telescope of double the aperture ought to trap the quanta four times as frequently, but there should be no other difference. This, as Lorentz pointed out in 1906, is quite at variance with our everyday experience. When the light of a star passes through a telescope and impresses an image on a photographic plate, this image is not confined to a single molecule or to a close cluster of molecules as it would be if individual quanta left their marks like bullets on a target. An elaborate and extensive diffraction pattern is formed; the intensity of the pattern depends on the number of quanta, but its design depends on the diameter and also on the shape of the object glass. Moreover, the design does not bear any resemblance whatever to the 'trial and error' design which is observed on a target battered by bullets. It seems impossible to reconcile this with the hypothesis that quanta travel like bullets directly from one atom of the star to one molecule of the photographic plate."

The wave theory explains the above phenomena perfectly. It does not, however, explain the emission of electrons under the influence of X-rays, as is emphasized by the following quotation from Dr. Ellis:

" To take a definite case, suppose X-rays are incident on a plate of some material, then it is found that electrons are ejected from the plate with considerable velocities. The number of the electrons depends on the intensity of the X-rays and diminishes in the usual way as the plate is moved farther from the source of X-rays. The velocity or energy of each electron, however, does not vary, but depends only on the frequency of the X-rays. The electrons are found to have the same energy whether the material from which they come is close to the X-ray bulb or whether it is removed away to any distance.

" This is a result which is quite incompatible with the ordinary wave-theory of radiation, because as the distance from the source increases the radiation spreading out on all sides becomes weaker and weaker, the electric forces in the wave-front diminishing as the inverse square of the distance. The experimental result that the photo-electron always picks up the same amount of energy from the radiation could only be accounted for by giving it the power either to collect energy from a large volume or to collect energy for a long time. Both of these assumptions are unworkable, and the only conclusion is that the radiated energy must be localized in small bundles."

It appears, therefore, that we require two different and irreconcilable theories of the propagation of light. The necessity for the quantum theory is not, however, universally admitted. Sir J. H. Jeans does not consider the hypothesis of light-quanta to be necessary, and the hypothesis that quantum phenomena involve essentially discontinuous processes is being attacked from more than one direction. An attempt has been made by Professor L. V. King to derive quantum phenomena in the atom from classical principles without introducing discontinuity, and in Schrödinger's theory, also, quantum conditions are obtained without violating continuity. The whole question, however, still presents great difficulties.

In this dilemma we may expect that very extraordinary hypotheses will be proposed. A very bold one has been suggested by Professor G. N. Lewis, who suggests that two atoms connected by a light ray may be regarded as in actual physical contact. The " interval " between two ends of a light ray is, on the theory of relativity, zero, and Professor Lewis suggests that this fact should be taken seriously. On this theory light is not propagated at all. This idea is in conformity with the principle that none but observable factors should be used in construct-ing a scientific theory, for we can certainly never observe the passage of light in empty space. We are only aware of light when it encounters matter. Light which never encounters matter is purely hypothetical. If we do not make that hypothesis, then there is no empty space. On Professor Lewis's theory, when we observe a distant star, our eye as truly makes physical contact with that star as our finger makes contact with a table when we press it. Whether this idea can be carried through re-mains to be seen, but it is symptomatic of the astonishing revolution in thought that we may expect.

Another question of some interest, and that cannot yet be definitely answered, is whether the fundamental pro-cesses of Nature are reversible or irreversible. As we saw in a previous chapter, science has hitherto assumed that fundamental processes are reversible. This conception, however, may prove to be inadequate. We may have to postulate an asymmetry in our causal laws, so that we may not suppose that processes may work backwards as well as forwards. As we have seen, the description of atomic phenomena seems to necessitate a reference to the final as well as to the initial state. Sommerfeld suggests, therefore, that the laws of the new quantum mechanics may be teleological, and that the old scientific notion of causality cannot be applied to them.

The enterprise of describing Nature mathematically, as we have presented it in this volume, has achieved some remarkable successes, but the present state of uncertainty in physics warns us that its final success must not be taken for granted. For the mathematical description of Nature to be successful it is absolutely necessary that natural events should be uniquely determinable, whether the determination involves reference only to an initial state, or whether it involves reference both to an initial and a final state. Science has long supposed that only an initial state was necessary. There are now some grounds for supposing that both an initial and a final state are required. In any case, for science to be possible, the data must determine what happens uniquely. Whether, in fact, physical phenomena possess this degree of determinateness may conceivably be doubted. In view, however, of the success that has hitherto been achieved by the scientific adventure, the doubt is somewhat academic. Perhaps an eighteenth-century mathematician, such as Laplace, would feel that modern physics had not fulfilled his hopes. Certainly Nature is more mysterious than the pioneers of this great enterprise suspected. But their ideal of showing that all Nature is a scientific unity, completely determinate, and obeying exact mathematical laws, is still the ideal of science. And it is the faith that this ideal may yet be realized which is the guiding motive of scientific men.

CHAPTER XII

GENERAL CONCLUSIONS

THE majority of civilized people to-day would certainly regard science as one of the major human activities. There are people, it is true, including some artists and religious persons, who regard science as unimportant or else as positively harmful. The vital problems of mankind, they point out, are spiritual problems. What is the good life? What is the significance of suffering? What is the chief end of man? These are the problems of most concern to mankind, and only those who have helped towards their solution are to be regarded as great teachers. Towards the solution of these major questions science, we are told, contributes nothing of importance.

It must be admitted that there is considerable justification for this attitude. The greater importance that men attribute to art and religion is not due simply to their ignorance of science. Art and religion satisfy deeper needs; the problems they deal with are intrinsically more important. But even if we say, with Einstein, that some great work of literature or music matters more to us than any scientific theory, the humanistic values of science are yet very considerable. In the first place it satisfies, within a limited region, the immensely important human need for comprehension. This is the chief justification for science. Scientific men wish to *know*, merely for the sake of knowing. That their discoveries may benefit the world, in some tangible material way, is to them a matter of relatively little concern. Very few discoveries of any importance have been made for the sake of their applications. Maxwell did not work at his electromagnetic theory of light with the object of providing every home with a wireless set. Those who share this appetite to understand find science, in spite of its limitations, irresistibly attractive. For, although the exact sciences deal with so limited a field of experience, the sort of understanding they give is of unequalled clearness and precision. Contrasted with them most other departments of thought appear vague and confused. In the modern world, indeed, where clarity and truth are so systematically dishonoured in the service of other interests, such a science as physics appears as a refuge, as a region where certain intensely congenial ideals are most clearly exemplified. In their private lives scientific men are, for the most part, as prejudiced and slovenly in their thinking as are most other men. But their science is the most completely honest of human monuments. Success is granted on no other terms. For its provision of an ideal of comprehension and for its insistence on accuracy science must be regarded as of incalculable humanistic importance.

Besides satisfying the desire for comprehension, science also serves the needs of practice, and this is the aspect of it that is usually taken as its chief justification. But the applications of science are merely means to ends, and the value of the means is to be judged in relation to the value of the ends. Science can help men to achieve what they want to achieve, whether it be saving life or destroying it. Certainly modern civilization would be utterly impossible without science, and it may be fairly argued that modern civilized communities are more desirable societies than have ever existed before. Possibly the strongest objection that could be made to our scientific civilization is that it tends more and more to produce the standardized man. But this lamentable fact does not seem to be a necessary consequence of the existence of scientific knowledge, and therefore, although scientific knowledge makes it possible, it is not to be blamed for it.

Another, and perhaps the most interesting, of the humanistic aspects of science lies in its relation to the rest of our knowledge and beliefs. What are the philosophical, and even the religious, implications of science? For the greater part of its history the most significant feature of modern science, philosophically speaking, is that it does not include values in its description of the universe. The mediaeval conception of purpose, applied to the operations of Nature, was found to be unnecessary so far as the science of physics was concerned. The ideal aim of physics, as we have seen, was to explain all phenomena in terms of masses moving through space and time in accordance with mathematically definable laws. But in the enunciation of these laws it was found unnecessary to refer to any ultimate purpose, and all the moral and aesthetic aspects of phenomena were regarded as irrelevant to the scientific description. Thus the persuasion arose that science supported materialism, the doctrine that the only objectively existing reality is a large number, possibly infinite, of little particles of matter whose purposeless combination produces everything, including our knowledge of the little particles.

This doctrine, as a philosophy, has always been singularly elusive, and many good minds have found it strictly unintelligible. But whatever standing it may have, regarded purely as a comprehensive and consistent theory, its experimental evidence has always been extremely slight. The theory that phenomena can be explained in terms of the motions of little pieces of " substance " was fairly successful in various branches of physics until the electromagnetic theory of light proved recalcitrant to this sort of analysis. But the extension of the theory to the whole of phenomena was, from the scientific point of view, little more than an idle speculation. To suppose, e.g., that the Ninth Symphony was produced by the random collisions of little hard particles was never more, from the experimental point of view, than a pleasing fancy. But although consistent materialists were probably always rare, the humanistically important fact remained that science did not find it necessary to include values in its description of the universe. For it appeared that science, in spite of this omission, formed a closed system. If values form an integral part of reality it seems strange that science should be able to give a consistent description of phenomena which ignores them.

At the present time this difficulty is being met in two ways. On the one hand it is pointed out that science remains within its own domain by the device of cyclic definition, that is to say, the abstractions with which it begins are all it ever talks about. It makes no fresh contacts with reality and therefore never encounters any possibly disturbing factors. This point of view is derived from the theory of relativity, particularly from the form of presentation adopted by Eddington. This theory forms a closed circle. The primary terms of the theory, " pointevents," " potentials," " matter," lie at various points on the circumference of the circle. We may start at any point and go round the circle, that is, from any one of these terms we can deduce the others. The primary entities of the theory are defined in terms of one another. In the course of this exercise we derive the laws of Nature studied in physics. At a certain point in the chain of deductions, at " matter," for example, we judge that we are talking about something which is an objective concrete embodiment of our abstractions. But matter, as it occurs in physics, is no more than a particular set of abstractions, and our subsequent reasoning is concerned only with these abstractions. Such other characteristics as the objective reality may possess never enter our scheme. But the set of abstractions called matter in relativity theory do not seem to be adequate to the whole of our scientific knowledge of matter. There remain quantum phenomena. These phenomena have not, so far, been derived from relativity theory, although some recent work makes it not impossible that they will be so derived.

Another way of meeting the difficulty mentioned is to assert that values must, after all, be included in the scientific scheme. It may be that some of the present difficulties of physics are due to the inadequacy of its primary abstractions. Professor Whitehead declares that the whole conception of " substance " possessing " simple location " in space and time is inadequate to the needs of modern science. On his theory time must not be regarded as pure succession. A moment of time mirrors in itself both its past and its future. According to Whitehead we directly perceive what he calls a " relationship of conformation " in the world. Things do not " simply occur "; they conform to previous occurrences. The notion of pure succession is an abstraction from the perceived relation of conformation. As he says [1]:

" In practice we never doubt the fact of the conformation of the present to the immediate past. It belongs to the ultimate texture of experience, with the same evidence as does presentational immediacy. The present fact is luminously the outcome from its predecessors, one quarter of a second ago. Unsuspected factors may have intervened; dynamite may have exploded. But, however that may be, the present event issues subject to the limitations laid upon it by the actual nature of the immediate past. If dynamite explodes, then present fact is that issue from the past which is consistent with dynamite exploding. Further, we unhesitatingly argue back-

[1] A. N. Whitehead: *Symbolism.*

207

wards to the inference, that the complete analysis of the past must disclose in it those factors which provide the conditions for the present. If dynamite be now exploding, then in the immediate past there was a charge of dynamite unexploded."

This notion is part of a doctrine that may be called a doctrine of universal relativity. Every volume of space makes reference to every other volume of space, and every moment of time to every other moment of time. The notion of matter is replaced by the notion of organism. Thus an electron is an organism, where an organism is defined as " the realization of a definite shape of value." Nothing is isolated; everything makes reference to everything else. Every event by reason of its very nature requires the whole universe in order to be itself. The theory is still in parts obscure, but it does meet certain difficulties, and it may well be that so comprehensive a re-organization of scientific abstractions will prove to be necessary. On this theory the notion of " value " seems to be fundamental. Values, moral and aesthetic, are regarded as integral constituents of reality, and the abstractions that led to materialism, even in their recent modified forms, are regarded as inadequate, not only for philosophy, but also for science.

The notion of an electron as an organism raises the question whether, in the reconstruction of the abstractions of physics, it will be found useful to borrow concepts from biology. It was implicit in the ideal of formulating all phenomena mathematically that biological phenomena would prove explicable in terms of physical concepts. There are not, at present, any convincing reasons for supposing this to be true, and a theory has arisen, the theory of " emergence," which denies the possibility. According to this theory the properties of a whole cannot always be deduced from the properties of its constituents. The reference here is not to merely technical difficulties of calculation, but to an intrinsic impossibility. The assertion is that supposing we know all the properties of an atom of chlorine and of an atom of sodium, taken in isolation, then it does not follow that from these properties we should be able to deduce the properties of a molecule of sodium chloride. New properties, it is asserted, may " emerge." On this theory chemistry can never be reduced to physics; the phenomena of life and mind can never be reduced to chemistry. At various stages of complication entirely new and unpredictable properties emerge. It is doubtful whether this theory could ever be proved, since we could always suppose our knowledge of the constituents of any given whole to be incomplete. But it is certainly a possible interpretation of the evidence that has so far been accumulated. Nobody has yet explained the properties of a living cell in terms of its constituent molecules, and even the properties of chemical molecules have not yet been explained in terms of the physicist's atom. But our belief or disbelief in the theory of emergence must rest, at present, on philosophical grounds.

Until the science of physics has overcome its present fundamental difficulties, it is premature to study in any detail what we may suppose to be its philosophical implications. Thus if physics finds it necessary to assume that space and time are discrete, the fact will doubtless be of importance to the philosopher. But it is by no means clear yet that physics will make that assumption. The notion of discontinuity may be found to be, after all, unnecessary in the regions where it is at present employed. But there are certain conclusions, of philosophical importance, that have been made more obvious by recent developments in physics. Perhaps the most important of these conclusions is that science deals wholly

209

with structure and not with substance. The raw material of science is our percepts. Corresponding to each percept we assign a physical cause. Different percepts, *e.g.*, red and blue, correspond to different physical causes. But the only properties of these causes that are investigated by science are their structural properties. Of the intrinsic nature of these causes science tells us nothing. Science mentions nothing in its analysis of the causes of red and blue (waves in the ether, or what not) that is analogous to the difference in *quality* of red and blue. It talks wholly about differences in structure, as by saying that one wave-length is longer than another.

But from a knowledge of structure we cannot deduce anything about the nature of the material that possesses the structure. The physical causes of our percepts are composed of events, and all we know about any physical cause is the structural organization of the events composing it. As to the essential nature of the events, " what they are made of," science tells us nothing. It seems probable, however, that these events are of the same nature as our percepts, that is, that they are what we call mental. A percept, if we are to believe the scientific account, is the last term of a connected process. The vibrating atom, the light wave, the process in the eye, the current along the optic nerve, the changes in the brain, the perception of the colour red, all appear as parts of a continuous chain. The one part of the chain that we know directly is the percept. The other parts are inferred. It seems natural to suppose that all parts of the chain are qualitatively continuous. In that case we may take the element we know most intimately, the percept, as giving us a clue to the constitution of the whole. It is quite possible, therefore, that the actual substance of the universe is mental, that the stuff of events is similar to percepts. The fact that a piece of matter has been reduced by relativity theory to a system of events, that it is no longer regarded as the enduring stuff of the world, makes the hypothesis that the " physical " and the " mental " are essentially similar very possible.

To conclude, we may say that science tells us much less about the universe than we used to suppose. It is limited, not only because, as an historical fact, mathematical laws have been formulated for only a limited class of phenomena, but because science, by its very nature, can tell us nothing about phenomena but their structure. Also, those elements of our experience that are ignored by science are not thereby shown to have no bearing upon the nature of reality. Our aesthetic and religious experiences need not lose the significance they appear to have merely because they are not taken into account in the scientific scheme. It is even possible that they will not always remain excluded from the scientific scheme.

APPENDIX

Arrangement of the elements in groups in order of their Atomic Numbers

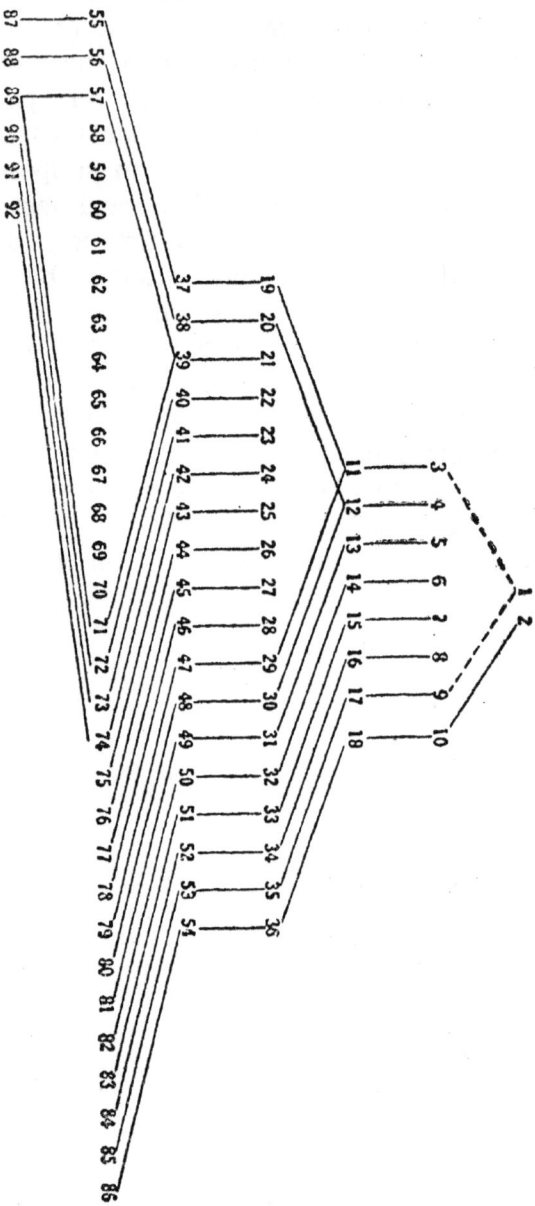

Table showing the groups in the Periodic System and which elements are related to one another in the different groups

THE BASES OF MODERN SCIENCE

No.	Name.	Weight.	No.	Name.	Weight.
1.	Hydrogen	1·008	47.	Silver	107·88
2.	Helium	4	48.	Cadmium	112·4
3.	Lithium	6·94	49.	Indium	114·8
4.	Beryllium	9·1	50.	Tin	118·7
5.	Boron	10·9	51.	Antimony	120·1
6.	Carbon	12	52.	Tellurium	127·5
7.	Nitrogen	14·01	53.	Iodine	126·92
8.	Oxygen	16	54.	Xenon	130·2
9.	Fluorine	19	55.	Cæsium	132·8
10.	Neon	20·2	56.	Barium	137·37
11.	Sodium	23	57.	Lanthanum	139
12.	Magnesium	24·3	58.	Cerium	140·2
13.	Aluminium	27·1	59.	Praseodymium	140·6
14.	Silicon	28·3	60.	Neodymium	144·3
15.	Phosphorus	31	61.	Illinium	—
16.	Sulphur	32·06	62.	Samarium	150·4
17.	Chlorine	35·456	63.	Europium	152
18.	Argon	39·9	64.	Gadolinium	157·3
19.	Potassium	39·1	65.	Terbium	159·2
20.	Calcium	40·07	66.	Dysprosium	162·5
21.	Scandium	44·5	67.	Holmium	163·5
22.	Titanium	48·1	68.	Erbium	167·7
23.	Vanadium	51	69.	Thulium	168·5
24.	Chromium	52	70.	Ytterbium	172
25.	Manganese	55	71.	Lutecium	174
26.	Iron	55·8	72.	Hafnium	—
27.	Cobalt	58·97	73.	Tantalum	181
28.	Nickel	58·68	74.	Tungsten	184
29.	Copper	63·6	75.	Rhenium	—
30.	Zinc	65·4	76.	Osmium	191
31.	Gallium	70·1	77.	Iridium	193·1
32.	Germanium	72·5	78.	Platinum	195
33.	Arsenic	74·96	79.	Gold	197·2
34.	Selenium	79·2	80.	Mercury	200·5
35.	Bromine	79·9	81.	Thallium	204
36.	Krypton	82·9	82.	Lead	207·2
37.	Rubidium	85·45	83.	Bismuth	208
38.	Strontium	87·63	84.	Polonium	210
39.	Yttrium	88·7	85.	Unknown	—
40.	Zirconium	90·6	86.	Radon	222
41.	Niobium	93·5	87.	Unknown	—
42.	Molybdenum	90	88.	Radium	226·4
43.	Masurium	—	89.	Actinium	(226–227)
44.	Ruthenium	101·7	90.	Thorium	232·1
45.	Rhodium	102·9	91.	Protoactinium	—
46.	Palladium	106·7	92.	Uranium	238·5